Praise for **JAMES SCHEFTER**'s | **The Race**

"Thrilling. . . . Some reporters get to cover the story of a generation. Count Jim Schefter among this lucky bunch."
—*Rocky Mountain News*

"Describes in great detail those tense, sometimes bitter days when the space race was based on one-upmanship, and beating the other guy was national policy."
—*The San Diego Union-Tribune*

"Brimful of racy incident and nicely written."
—*Mail on Sunday* (London)

"Every journalist deserves at least one epic story. For James Schefter . . . it has been humanity's outreach to interplanetary space."
—*The Christian Science Monitor*

"Dramatic and entertaining. . . . Shot through with sexual energy."
                                                    —*The Economist*

"Schefter's familiarity with NASA's top men is the greatest strength of the book. . . . It belongs in the library of any serious space enthusiast."
                                                    —*St. Petersburg Times*

"Laced with funny details. . . . Takes readers close to the early days of America's space program, its tragedies and its triumphs."
                                                    —*The Daily Oklahoman*

JAMES SCHEFTER

The Race

James Schefter covered NASA from 1963 to 1973, for the *Houston Chronicle* until 1965, and then for Time-Life. His work has appeared in magazines as diverse as *Popular Science, Reader's Digest,* and *Paris Match.* Schefter is the author of *All Corvettes Are Red: The Rebirth of an American Legend.*

THE
Race

*The Complete True Story
of How America Beat
Russia to the Moon*

JAMES SCHEFTER

ANCHOR BOOKS

A DIVISION OF RANDOM HOUSE, INC.
NEW YORK

This is for my son Mike, who came to Earth while Neil and Buzz were still on the moon, and through the fluke of retardation will never understand. And for his younger brothers, Will and Drew, strong and smart on their own, and better men because of Mike's innocent example.

FIRST ANCHOR BOOKS EDITION, JUNE 2000

Copyright © 1999 by James L. Schefter

All rights reserved under International and Pan-American Copyright Conventions. Published in the United States by Anchor Books, a division of Random House, Inc., New York, and simultaneously in Canada by Random House of Canada Limited, Toronto. Originally published in hardcover in the United States by Doubleday, a division of Random House, Inc., New York, in 1999.

ANCHOR BOOKS and colophon are registered trademarks of Random House, Inc.

The Library of Congress has cataloged the Doubleday edition as follows:
Schefter, James L.
The race: the uncensored story of how America beat Russia to the moon / by James Schefter. —1st ed.
p.   cm.
Includes index.
ISBN 0-385-49253-7
1. Astronautics—United States—History. 2. Space flight to the moon—History. 3. Project Apollo (U.S.)—History. 4. Space race—History. I. Title.
TL789.8.U5S34   1999
625.45'4'0973—dc21   98-54430
CIP

Anchor ISBN: 0-385-49254-5

Book design by James Sinclair
Author photograph © Phil Dean

www.anchorbooks.com

Printed in the United States of America
10 9 8 7 6 5 4 3 2 1

Preface

Four high school seniors sat in a black 1951 Dodge parked at Phil and Lil's Drive-In in Grand Forks, North Dakota. It was chilly, the hint of winter already on the winds blowing down from Canada. Even the lights of the small town, population 24,000, couldn't wash out the stars. Saturday night, October 5, 1957, had come on a tail of fire that was pushing our world into a new and amazing era.

While three of the boys fiddled with the radio and waited for hot, greasy cheeseburgers and fries, one ran next door to the Dairy Queen. He was dating the owner's daughter and she was working that night. That should be good for free shakes all around. Back and forth across the AM band, the three in the car listened for news. It was one of the few times in their teen years on the remote Dakota plains that rock and roll was not just ignored but unwanted from the car radio's single speaker.

Something beyond the sounds of music and the urges of hormones dominated the boys in the car. It was, in fact, something beyond Earth itself. The Russians—*the Russians!*—had put an artificial satellite into orbit around the world the night before. Suddenly the vastness of the plains, stretching far to the west until stopped by mountains and running north beyond Winnipeg, maybe all the way to Hudson's Bay, didn't seem so great. The Great Plains of their ken was dwarfed by a 184-pound shiny sphere that could do nothing but broadcast an irritating beep from its radio out there in space.

They pulled in a Canadian station, CKY-Winnipeg, and listened to the news.

"Jesus, they did it," muttered Fred Johnson. He would go on to get a

Ph.D. in advanced mathematics and become a respected government scientist in future decades, influencing policies that he could not imagine at age sixteen.

"Fucking Russians," mused Morley Glicken, whose one-day Ph.D. in social work would put him face-to-face with the poor and the dysfunctional in a society driven by technologies that did not yet exist. His father walked out of Russia forty years earlier; the family memories of despair and hunger were still vivid.

Ordeen Flom crawled into the backseat, a single DQ malt in hand. He ignored the glares of his friends, and shrugged. "Do you suppose we can see it?" After college, he would experience battle in Vietnam, end his Army service as a major, and ultimately settle down as a Denver businessman.

"The radio says maybe, with binoculars," I answered. I was seventeen and would, in a few short years, become the resident correspondent for both *Time* and *Life* magazines at the yet-to-be conceived Manned Spacecraft Center in Houston.

The memory of that night still burns in my mind. We sat in my old black Dodge and talked about things we had no way of understanding. Flash Gordon was suddenly real, and how could this science nonfiction fit into the only lives we had known, in an isolated town where television arrived only a few years before? When we saw pictures of other places or great events, they were in *Life* or *Look* or *National Geographic* or in the movie newsreel. The radio was our link to the world, but we had grown beyond the years of listening to *The Shadow, The Green Hornet, The FBI in Peace and War,* and *The Lone Ranger.* Our imaginations had been trained and honed by radio, but now the radio was mostly used to listen to Elvis and Pat Boone. The Everly Brothers were new kids on the scene, Conway Twitty had yet to invent himself, and the Russians—*the Russians!*—had beaten us into space.

It was a pivotal moment in the history of the world. Even we four callow and unsophisticated boys of the prairie saw this truth dimly against the brilliant stars set into a black October sky. We didn't see Sputnik, but however blurred our vision, we did see the future. That weekend changed all of our lives, as it changed the world we would live in. And it set my own course, opening the way to many adventures and to an emotional ride through great events. In barely six years, at age twenty-three, I would be handed a front-row ticket to the space race, and nothing and no one could make me relinquish it.

I would know and rub shoulders with the great and the famous of the space race, at least on the American side. I would see the reality of sudden death when astronauts gave their lives in pursuit of the finish line. With the special entrée granted to *Time* and *Life,* I would see things inside the manned space program denied to other journalists. When NASA finally gave journalists permission to enter mission control during a manned space flight, I would be selected by my peers in the world's press as the print correspondent to be called in an emergency. That mission was Apollo 13 and the *Flash* message sent around the world by the wire services that there would be no moon landing came from my pen. By then, of course, the race was long over.

I aided and abetted in successful conspiracies to smuggle items to the moon; I loaned my plane to astronauts and walked with them through the jungles of Panama, the volcanic interior of Iceland, and the stark terrain of the California Mojave. I saw. I knew. I recorded. I remembered.

For most of *The Race,* and its aftermath, I was there. I saved every notebook, every written report sent to Time-Life in New York and on a freelance basis to other publications, and cases of NASA documents and photos. I arrived at the end of the Mercury program, but soaked up stories of the years when NASA was an infant, then a toddler staggering toward its grand destiny. For more than a year in doing this book, I talked to the still-living from those days, dug into official and unofficial archives of papers left by the living and the dead, and brought the story of the space race back together in ways that have not been told until now.

At the end, this is a story cloaked in history, or perhaps history cloaked in a story. It is not my story, except that I am the one who gets to tell it. This is a different side of something familiar. It is the gut-level story of courageous, inventive, and driven people who accomplished the greatest feat of the twentieth century first by conquering space, then by landing men on the moon. In the end, it is an American story with Russian sidebars.

Finally, I hope, *The Race* is just a darned good story that happens to be true.

Acknowledgments

The Race is not a scholarly work, but it benefited from the advice and assistance of scholars. Foremost among these are Sergei Khrushchev, amazingly ensconced as a professor at Brown University, and Roger Launius, NASA's eminent and skilled historian. Sergei's stories of life with his father and Roger's keen eye in reading the manuscript added much to this book.

Christopher Columbus Kraft, Jr., was generous in his time with me, and when he allowed me access to his private papers at Virginia Polytechnic University, he opened a treasure trove of details, secrets, and insights. As I read his files, my most frequent thought was "If I'd known this in 1966 . . ."

Some who helped are adequately referenced in this book. Their stories are part of *The Race*. There are too many others to name, but I hope they'll recognize their contributions if they read these pages.

Three important people must be mentioned. The memories and stories told to me by John Glenn and Scott Carpenter were wondrous and enlightening. Then there's *Life* photographer Ralph Morse, old friend and slave driver, called back into harness to turn his camera again on John Glenn's amazing space shuttle flight in 1998. He is as vital and active at eighty-one as he was when I first worked with him in 1965. His recollections gave *The Race* both insights and chuckles that would otherwise be missing.

Kat James was stalwart in supporting me in the up days and the down days, and suffered through many readings and revisions as *The Race* took form. She has a discerning eye, and not just as a reader. Doug Williams and

Beverly Gail did their usual, saying three important words often and loudly: "Go get 'em!"

My agent, Dominick Abel, was solid in believing in *The Race,* and in finding my editor, Bill Thomas of Doubleday. Bill knew when to let me run and when to pull on the reins; he made me better than I am.

John Stewart was most gracious in granting permission to quote lines from his wonderful 1969 song, "Armstrong."

The analysis and conclusions about the Russian space program are mine, as are the choices about the men most important in running this race. I believe these are both new and accurate. Any factual mistake here is also mine. I hope they number between few and none.

To all who helped, I can only say thank you. I hope you enjoy reading this book as much as I enjoyed writing it.

Park City, Utah
January 1999

the world all stopped to watch it, on a July afternoon,
when a man named Armstrong walked upon the moon

—John Stewart, "Armstrong"

THE
Race

The Finish Line

He stood awkwardly, looking out the window at a gray and dusty landscape. It had been a long trip, either four days or fourteen years depending on your perspective. Now it was time to finish it. In the next five minutes, he had to go out there. Then he had to make a comment for the record.

I have no idea what to say.

He raised his eyes to the horizon only six miles away. Between here and there, nothing changed much. *But between* there *and* here, *everything has changed. How did we get from* there *to* here?

He looked down to the ground and it didn't seem so far away. *It's just a small . . .* His mind took the thought and raced to the logical conclusion. *Of course! It's both. It's always been both. That's what I should say.*

With a nod to his partner and an unseen smile, he turned around, crouched, and began to back out. The porch sloped down just a bit. The steps were easy, but the last one was higher than he'd expected. He jumped into the big circular dish at the bottom and still hadn't touched ground.

An entire world watched grainy black-and-white television images of his progress. Listened to him briefly describe what he saw. Wished he'd get on with it.

"Okay." His voice was scratchy with distance and static. "I'm going to step off the LM now."

He put his left foot down into the dust. Felt that the ground was solid underneath. Took a deep breath. And spoke to history.

"That's one small step for man . . ."

1

Nobody believed Leo Sedov.

It seemed absurd, talking about Russian rockets and space travel, but there he was, a senior Soviet scientist calling a press conference in Copenhagen to make yet another absurd claim about his country's technological greatness. Such claims were so common in the 1950s that even the Sunday funnies parodied them. Americans were accustomed to reading claims that Russians had invented the bicycle, the electric lightbulb, the telephone.

Americans laughed at the peasant Russians and understood that only a backward people with a great inferiority complex could be so silly. Hadn't America only recently provided Russia with the technology and hardware they lacked, so they could help win World War II? And then the ungrateful louts had responded with coarseness, spitting in the American face by blockading Berlin, by creating an iron curtain across Europe, by pretending to be better than they were.

Now this. The Sixth Congress of the International Astronautical Federation was convened in Copenhagen, and for the first time, the upstart Russians were there, if only as observers. The important people, the *participants,* were from the Western and developed countries. But here stood Academician Leonid I. Sedov, a Soviet physicist, telling fifty or so reporters at the Soviet embassy that the Soviet Union would soon put an artificial satellite into orbit around Earth.

He didn't tell them that the idea of a satellite was already old news in his country. Two experienced Russian rocket men had been pushing the idea for more than a year. One was Mikhail Tikhonravov, whose 1954 paper laid out technical details for a satellite. His latest revision of that paper was barely two weeks old when Sedov stood before the press in Copenhagen. The other was a mysterious figure known only as the Chief Designer. An old friend of Tikhonravov's, he would carry the satellite idea into the corridors of Soviet power and place it before commissions, department ministers, and even Premier Nikita Khrushchev himself. None of this was known to westerners when Sedov spoke. Seeing the skeptical looks in the audience, particularly on the faces of the American reporters, Sedov said it again even more force-fully: "The realization of the Soviet project can be expected in the near future."

The date was August 2, 1955.

Only four days earlier, James C. Hagerty, press secretary to President Dwight Eisenhower, had made a similar announcement in Washington. The United States, Hagerty said, would break out into space with "small Earth-circling satellites" during the eighteen-month-long International Geophysi-cal Year (IGY) that would begin July 1, 1957. It was a fantastic proposal, but coming from Americans, at least it didn't sound crazy. Rocket technology in the mid-1950s was almost good enough to reach space. If it could get just a little better—able to push a small payload to 17,500 miles an hour—that would be fast enough to overcome gravity.

The IGY had been years in planning, an international study of the upper atmosphere and outer space with high-altitude balloons, sounding rockets, ground-based telescopes, and other instruments. An IGY committee ap-proved an American-sponsored proposal in October 1954 to include satel-lites on the research list, but until the White House announcement nearly ten months later, no country had stepped up to that particular challenge.

Now the Russians—*the Russians!*—were going into space too, and in the "near future" no less. The *New York Times* gave the story fewer than four hundred words on an inside page, under a one-column headline reading

SOVIET PLANNING
EARLY SATELLITE

Russian Expert in Denmark
Says Success in 2 Years
Is 'Quite Possible'

Across the broad stretch of America, where small-town dailies, not the mighty *Times,* were the choice of readers, and where television had yet to make a major impact, the story was a nonstarter. For all the serious attention he got in the United States, Professor Sedov might as well have made his prediction in the Sunday funnies. Certainly more Americans would have seen it. But except for a few scientists in the inner circle of U.S. rocketry, none would have believed it.

Nor would the average American have believed the decision reached twenty-eight days later in Moscow. The shadowy Chief Designer convinced his peers in the Soviet Academy of Sciences to create a commission that would oversee launch of an orbiting satellite *before* the beginning of the IGY. The Chief Designer's timing had only one purpose: to beat the United States of America into space. It still needed official government approval and that would come in a matter of months. But still, there it was. The space race that would change the face of science, technology, human spirit, and international politics was announced in a public press conference on August 2 and *de facto* began on August 30, 1955.

Only one of the two participants heard the starter's pistol.

The last thing President Eisenhower wanted was any kind of race with the Russians. Some of his advisers thought differently, but Ike was the boss.

Historians agree on those points. They disagree about why Ike was so adamant. Each round of declassified national security documents from the late 1950s to the mid-1960s gives rise to new interpretations about Ike's inexplicable blindness to the terrible impact of losing a race he may not even have known he was running. A word, a wink, a nod from the president of the United States on the subject of quickly putting a satellite, *any satellite,* into orbit around Earth would have changed the course of mid-twentieth-century history and avoided the worst crisis of his presidency.

But still, Ike was warned. Even Harry Truman had been warned. Way back in 1946, when German rocket parts and Wernher von Braun's rocket team were in exile-residence at the Army's Fort Bliss, by El Paso, Texas, a RAND Corporation report called it right: "The achievement of a satellite craft by the United States would inflame the imagination of mankind, and would probably produce repercussions in the world comparable to the explosion of the atomic bomb."

World War II had just ended and nobody in power cared. Von Braun, of course, cared, but he was still a captured German scientist yet to prove himself to his new American masters. At a backwater research facility in Langley, Virginia, a few others dared to think about spaceflight too. But they had their hands full with advanced research on aeronautics and upper-atmosphere phenomena, so they didn't think about it much. Their names would become legendary in the race Eisenhower wasn't running. Bob Gilruth. Max Faget. Chris Kraft. But that came much later.

Seven years after the RAND report, an atomic bomb scientist named Aristid Grosse prepared another pro-space report. It was 1953, and for several years now, Americans had been in the process of being conditioned to accept space travel as a possibility, if not yet reality. Science fiction writers like Ray Bradbury, Robert Heinlein, and Arthur Clarke were hitting their stride with popular novels about satellites, moon travel, and adventures in outer space. In their books and stories, space was mostly conquered by Americans, or people who mostly traced their ancestry to Americans. Hollywood did its share too, with movies like the 1950 *Rocketship X-M,* a not-bad moon–Mars journey starring Lloyd Bridges and Hugh O'Brian that was far more believable than the earlier Flash Gordon Saturday serials, and *Destination Moon* in the same year. Heinlein coauthored *Moon*'s script, and its special effects won an Academy Award. More important, *Destination Moon* inflamed the imaginations of teens and preteens. A nine-year-old in 1950 would be the average adult working in a certain windowless room in Houston nineteen years later.

If more Americans had known how close they were to making orbital flight happen, the space race might have run a different course. Aristid Grosse of Temple University knew. His keen perceptions in a commissioned report landed on the Pentagon desk of Don Quarles, whose title was Assistant Secretary of Defense for Research and Development.

Grosse seemed to understand the Russian psyche. They'd only been able to develop their own atomic and hydrogen bombs by stealing the basic technology from the United States. They knew it. The United States knew it. And they knew the United States knew it. Now they yearned to be first in something, to prove something to themselves and to us. Launching a satellite might be it. The impact of a Russian conquest in space would be cataclysmic. Grosse thought they would try. "If the Soviet Union should accomplish this ahead of us, it would be a serious blow to the technical and engineering prestige of America the world over," Grosse wrote to Quarles. "It would be used by Soviet propaganda for all it's worth."

Quarles put the report into the F&F category. File and Forget. There's no clear record that Eisenhower ever saw it. In any government or corporate hierarchy, people like Quarles know that their job is not only to provide information to the boss but to keep some information from him.

Next up to the space race starting line, after another two years had passed, was Nelson Rockefeller. By now the United States had agreed, inside the government, to orbit a scientific satellite as part of the International Geophysical Year. Part of its reasoning was to establish that national sovereignty over its airspace was precisely that and no more. Where air ended, so did sovereignty. Nothing else would serve; as Earth turned below it, any satellite would cross over national boundaries. National or international law couldn't overrule the laws of physics.

Besides, that particular law of physics made a satellite perfect for observing our enemies. But therein lay another rub. Would the Russians cry foul, accuse the United States of illegal overflight, even challenge America as a warmonger for flaunting or brandishing its intercontinental missile technology? The skittish among Eisenhower's people saw an American victory in any space race as a propaganda coup for *Russia,* not for the United States. That kind of muddleheaded thinking wove its wicked threads into the internal discussions of space in 1955. In May, Rockefeller, in his role as a key Eisenhower adviser, vehemently disagreed. Letting the Russians go first into space would have "costly consequences," he said, adding that "this is a race that we cannot afford to lose."

Nobody listened to him either.

So the plans for an IGY satellite drifted forward to the announcement in late July that the United States would launch one after mid-1957 and before the end of 1958. Where future presidents would make such announcements themselves, surrounded by science advisers, congressmen and senators, and space professionals, Eisenhower left this one to his press secretary.

If they'd known how the Chief Designer would react, they might have played things differently.

The Chief Designer was Sergei Korolev. His name and face would soon be erased from Soviet scientific literature, including from membership lists in technical organizations and societies. When he was quoted in public, it was only as the Chief Designer. When he wrote for the press or authored a scientific paper, he used a pseudonym, Professor K. Sergeyev, a neat twist on his true identity. Korolev was tough. And he knew irony when he practiced it.

So on one level Korolev became a Soviet nonperson. The price of technological success in the Soviet Union of the 1950s and '60s was to disappear from public view. Potential fame required anonymity. If the West knew who he was, the reasoning went, secret agents might try to kidnap or kill him. But despite his anonymity, Sergei Korolev became a powerhouse in the Soviet space program, a man who could and did pick up the phone to call Nikita Khrushchev and who could and did humiliate the mighty United States of America.

Khrushchev was from peasant stock and was shrewd. When the Soviet premier was briefed on the progress of the intercontinental S-7 rocket, the Chief Designer took the opportunity to press for the highest possible approval for putting a satellite into space. Khrushchev's son Sergei, who would have a distinguished career in the coming space program, remembers that his father's only question was, "Will it hurt the ballistic missile program?"

"No, no," Korolev assured him, "we will use the same rocket, only changing what we put on top."

"Well, then you should go ahead and do it." Korolev had his marching orders.

The Chief Designer was born in the Ukraine in late 1906. His parents soon split, but he thrived under the tutelage of his grandparents and was an early reader. His natural father taught literature, but it was the engineer his mother married in 1916 who influenced the boy into the life he would lead. They moved to Odessa in time to see World War I incursions by Germans, British, French, Serbs, and more. The Bolshevik Revolution added to the keen-minded Korolev's learning stew, and by his midteens he was a student in a school that taught everything from physics and mathematics to carpentry and roof painting. Korolev soaked it up.

From Odessa he returned to school at Kiev Polytechnic, then two years later moved on to a science and engineering school in Moscow. He was not yet twenty and plunged into the study of aviation. One of the lecturers at Moscow Tech was Andrei Tupolev, who became an icon for his aircraft designs. Sergei Korolev listened avidly and learned. He graduated in 1930 after designing his own small airplane. Tupolev was his adviser.

It was natural that Korolev turned to rockets. He seemed to be a man always in search of a challenge. In rocketry, there was a Russian who *was* first. Konstantin Tsiolkovsky developed theories and principles for rocket flight and in 1903 calculated the speeds it would take to hurl free of gravity and go into orbit. His writings received wide attention inside Russia, and eventually throughout the world. Rockets powered by gunpowder or other burning solids had been around since the early Chinese invented them. But it took Tsiolkovsky to make the imaginative leap into space rockets and to integrate the mathematics and physical laws that would control them. He wouldn't be first to launch a liquid rocket, though. That breakthrough belonged to American Robert Goddard, whose liquid-fueled rocket reached an altitude of four hundred feet in 1926.

Korolev's regular job was in an aviation design bureau in Moscow. On weekends, he devoted his energies to a rocket research group called the Group for Studying Reaction Propulsion. His work contributed to a rocket engine, sophisticated for the time, that developed 110 pounds of thrust. It was far too small for spaceflight, but it could lead to some kind of rocket propulsion for an airplane. When they tried it on a glider, after fighting through numerous technical problems, the test gave Korolev a taste of defeat. The rocket burned through and that particular design was abandoned. It wouldn't be his last bitter moment in the world of rocketry, but it surely helped to prepare him for later events.

One of those life-altering events came in 1938. Korolev was summarily arrested, charged with impeding progress on a new rocket program, and eventually shipped off to Kolyma in the Siberian gulag. Paranoia was rampant in Russia under Joseph Stalin, and it took no more than accusations by jealous associates to destroy a man. Korolev was lucky. He survived.

After months of hard labor in 1939, Korolev was released for reinvestigation. Some say that Tupolev intervened, but since he himself was under arrest in a less-harsh prison, that may be apocryphal. Korolev was stranded in Siberia over a long winter, his health in ruins, his teeth lost to scurvy, his strength sapped. Yet he made it eventually to Moscow, where his case was reviewed. Again he was found guilty, but his sentence was reduced, from ten years to eight, and through intercessions of those who knew him, he found himself with Tupolev in relatively mild confinement in Moscow. Their prison, in fact, was an aviation design bureau where Tupolev, with Korolev at his side, developed the World War II Tu-2 bomber and other aircraft used in defeating Hitler's Nazi armies.

Korolev regained his health in Moscow, but the good times didn't last. With the Germans approaching, the entire Tupolev team was moved to Omsk, in Siberia, where design work continued. There was enough to eat, but it was still prison. As the work wound down, Korolev was moved again and again, ending up at a prison for rocket and jet engine development in the Caucasus where he became a chief engineer. By now Korolev had been a prisoner for more than six years. He had been able to work on airplanes and now on the most advanced of engine technologies for much of that time, but the effects on his body and his psyche would never go away. He was finally released in August 1944 and stayed on in Kazan until the following spring. Shortly thereafter, the inexplicable happened.

Sergei Korolev, aircraft designer, jet and rocket expert, long-term prisoner, was commissioned a colonel in the Red Army and sent to Germany. There his assignment was to look into the rocket work done by the German team led by Wernher von Braun. Von Braun himself, and most of his people, were already at Fort Bliss, Texas. But what they left behind was most informative. Korolev was impressed with German technology and with the success of the V-1 and V-2 rockets in raining terror on London. He was so impressed that, when he had the chance a few years later, he made certain that the German experts captured by Russia were so segregated in their Soviet-sponsored R&D that they could never be a threat to Korolev himself, or to the programs he would lead.

In the postwar years, a new side of Sergei Korolev emerged. Maybe it was influenced by his experience as a prisoner, because he grew autocratic and played the games of bureaucratic intrigue with increasing skill. He was a hard boss, and grew harder with age. He had been tested and survived. He would test those around him to the last days of his life.

Any serious money the United States spent on rocketry immediately after World War II went into creating newer and better missiles for the military. The lessons of von Braun and London had not been lost on American strategists. And von Braun himself was diligently working at Fort Bliss to demonstrate that the V-1 and the far more sophisticated V-2 were only the start.

Another of the German rocket men would not go to Fort Bliss. General Walter Dornberger, the man who ran the rocket program for Adolf Hitler, would eventually become a vice-president at Bell Aerosystems in Buffalo, New York. Dornberger hired the precocious von Braun ("He wasn't 'von' then," Dornberger would say, and never referred to Wernher as anything but Braun) from a Swiss school at age eighteen and nurtured his career under the Nazis. Dornberger made much of his own role in developing the V-series terror rockets. "I was having lunch with Hitler," he once said to a startled interviewer, "and he'd had a dream that we couldn't hit London. He stopped our budget and delayed us a year."

Ten thousand miles from Fort Bliss, Sergei Korolev was working too. Russia had its own captured Germans and they were teaching Korolev, his people, and a whole new generation of young Russian engineers about the intricacies of rocket-powered flight. What rockets and parts, machinery, and documentation the Americans hadn't hauled off to Texas, the Russians— starting even before the war ended and taking advantage of the fact that most of the German rocket development took place north and east in the country—packed up and moved even farther east into their homeland. Along with hardware, they eventually took thousands of technicians and the handful of senior scientists and engineers who hadn't gone west with von Braun. By late 1946, most found themselves and their families in settlements near Moscow. They would have familiar surroundings at work. The Russians stripped the Mittelwerk production facility where the V-2 was mass-produced, anything at the Peenemünde development center that had survived massive bombing, and the unscathed rocket engine center at Lehesten, and carted all of it off to the Soviet Union.

But until the decision to take it all home, the Russians in 1945 and '46 set up shop inside Germany to drain details from the minds of the Germans. When Korolev arrived in September 1945, much of the groundwork for co-opting German technology was in place. He focused on organizing the transfers and on seeking out and questioning the most knowledgeable Germans in highly technical terms. His own knowledge of the basics impressed the Germans, and he soaked up the technical details of designing, building, and launching the big V-2 rockets with encyclopedic ease. Even then he sometimes spoke of using this technology to put satellites into orbit or to go to the moon.

Korolev returned with the Germans in tow and was soon sent to a new rocketry institute with the acronym NII-88 (the translation of the full Russian name is Scientific Research Institute-88). About 150 Germans were assigned there with him, though only a few had been at Peenemünde with von Braun. At NII-88, Korolev became, for the first time, a chief designer, the equivalent in the United States of program manager. His program was a long-range military rocket, and in this new assignment he thrived.

In the years following, Korolev apparently masterminded a devious method of milking knowledge from the Germans while keeping them from becoming prominent in the Russian program. Young and inexperienced Russian engineers, some of them no more than students, worked side by side with German experts in assembling captured rockets and preparing them for tests. The senior Russian experts, including Korolev, stayed at arm's length from the process. But the young Russians were more than acolytes. They got the education and practical experience that Korolev wanted for them, and they were the intermediaries carrying highly technical questions, arguments, and observations between the two camps. It was cumbersome, but it ensured that the Germans would not get too much credit for work being done concurrently by the senior Russian experts.

When the Germans were asked to design an advanced rocket with a four-hundred-mile range and improved accuracy, Korolev set his team working on the same thing. And when the Germans produced such a design in late 1947, so did Korolev. The problem for the Soviet technical hierarchy was to decide between the two. The answer was to keep both groups working in 1948, then '49 and '50, the Germans still in isolation, with their brains being picked continuously. Korolev examined their reports avidly and finally concluded that he and his people knew as much as their Teutonic "teachers."

The German effort was suspended in 1950. Most of them were soon repatriated to East Germany. Years later a standing joke among American journalists covering the space race held that it was being run by Germans, "their Germans against our Germans." The stranger truth remained hidden for decades. A man nobody in the West knew, the mysterious Chief Designer, had shamelessly used "their Germans" to build the foundation of Soviet rocketry. But instead of co-opting them for the future contributions they might make, as America did with "our Germans," he shunned them, shut them off, and when his personal goals had been met, sent them packing.

That left Sergei Korolev in charge of Russia's destiny in space.

Through the mid-1950s, the Soviet Union and the United States of America were in a race, and the leaders of both countries knew it. It was a race to dominate the world, to decide whether American-style democracy or Russian-style communism would be Earth's predominant political philosophy. Neither side was winning.

Much of the race was run on the field of technology, and the technology of choice was weaponry. The United States led in long-range bombers and in atomic bombs to drop from them. The Soviet Union countered American technology with its men under arms and its massed military forces aimed at Western Europe. Both sides were developing intercontinental missiles and hydrogen bombs, with the U.S. holding a slim lead. The United States detonated the first hydrogen bomb in late 1952. Barely nine months later the Soviet Union exploded its own H-bomb. With the race tied on that front, it moved on to the realm of missiles. Big bombers carrying atomic or hydrogen bombs could be shot down. Incoming missiles couldn't.

Sergei Korolev had become Chief Designer of the Soviet long-range missile program, developing a rocket called the S-7. In the United States, responsibility was spread across the Army, Navy, and eventually the Air Force. The Army had a major asset. It owned Wernher von Braun and his team. The German rocket men and their families had been moved from Fort Bliss to the Army's Redstone Arsenal at Huntsville, Alabama. There they used their V-2 skills and technology to develop the Redstone missile, and later the multistage Jupiter rocket. Von Braun's day-to-day work on war rockets only consumed part of his energies. Even in his Peenemünde days, he thought about spaceflight. He wrote a detailed technical paper about a manned mission to Mars that was read at the Second International Congress on Astronautics in September 1951. Von Braun himself was not in attendance. The congress was in London and he was smart enough to stay away from the city upon which his rockets had rained so much misery.

But his paper, full of technical detail, created a stir. Congress Chairman Arthur C. Clarke was ebullient. "Spaceflight is likely to be the next major technical achievement of our species," he crowed. Across the Atlantic, *Time* magazine gave von Braun's paper rapt attention. Its caption on his photo tagged him "Astronaut von Braun," certainly one of the first times that appellation was used.

The next year *Collier's* magazine contracted with von Braun and other visionaries, including Willy Ley and astronomer Fred Singer, for a series of articles on man in space. Teaming with legendary artist Chesley Bonestell, von Braun set forth stirring descriptions of manned rockets, wheel-like space stations, and expeditions to the moon. Bonestell's paintings of winged spaceships, space station shuttle craft, and lunar landers appeared on the magazine's covers and spread across its inside pages. Three von Braun articles ran in 1952, further "educating" Americans about the coming space age. He told readers that a manned moon mission would last just five days and that "the project could be completed within the next 25 years. There are no problems involved to which we don't have the answers—or the ability to find them—right now."

Time gave its December 8, 1952, cover to a planetoid-strolling robot over the title question, "Space Pioneer. Will man outgrow the earth?" It recalled a Hayden Planetarium offer in 1950 to take reservations for visits to the moon and planets. The Hayden got 25,000 responses, leading *Time* to conclude in part that America was already "oversold" and by "mixing fact and fiction, apparently believes that space travel is just around the corner."

Some experts strongly disagreed. Dr. Milt Rosen of the Naval Research Lab, then head of the Navy's Viking program for high-altitude research rockets, was "frankly aghast at difficulties that von Braun lightly brushes aside." An unnamed critic was more blunt: "Look at this von Braun! He is the man who lost the war for Hitler. His V-2 was a great engineering achievement, but it had almost no military effect and it drained German brains and material from more practical weapons. Von Braun has always wanted to be the Columbus of space. He was thinking of spaceflight, not weapons, when he sold the V-2 to Hitler. He says so himself. He is still thinking of spaceflight, not weapons, and he is trying to sell the U.S. a spaceflight project disguised as a means of dominating the world."

Von Braun's reply aimed for the high ground. He called for "a daring, inspiring program that has a real chance of controlling the world . . . a satellite station [that] would put the U.S. far ahead in the race for power . . ." Almost in afterthought, *Time* wrote, "He hopes that the Russians will not be the first to garrison an orbit."

The next year in *Collier's,* writing with Cornelius Ryan, von Braun described a "baby space station," an orbiting ship shaped like an ice cream cone and carrying three rhesus monkeys as passengers precursing flights by men. The monkeys would die "in a long white streak of brilliant light" during the ship's destructive reentry into the atmosphere. Bonestell's painting of that moment is a classic in space art. There was more truth than anyone knew in the articles and in the paintings. While von Braun wrote majestically of spaceflight, he continued to push his team onward in solving the problems of uprating rockets from short and midrange achievement to intercontinental range and then on to outer space. On the other side of the world, Sergei Korolev drove his team toward the same goals.

Korolev and von Braun never met. But in 1953 they shared a dream.

The approaching International Geophysical Year threw the world's scientists together and they talked. It took only a few dinners before the Russians told the Americans exactly what was coming, and when.

Dr. John Simpson, who went on to become director of U.S. Fermi National Laboratory, was a member of the IGY organizing committee and a recognized expert in high-energy physics. The committee met in Barcelona in September 1956, with key scientists presenting their nation's contributions to IGY studies. Among the American presentations was a slick and glossy discussion, complete with colorful graphics, of the satellite that the United States planned to launch. It would not be a big satellite, a silvery sphere weighing only twenty-two pounds. But it would carry instruments to measure radiation and other phenomena beyond Earth's atmosphere.

Science was an important, even critical, factor in the Eisenhower administration's satellite plans. Ike's advisers were deathly afraid that America would be branded as a warmonger if it used a purely military rocket—an intercontinental ballistic missile, or ICBM—to put a satellite into space. The military would have to be involved, of course, because no civilian agency possessed rockets of any size. But science had to be the key to space.

Simpson and his fellow American scientists knew that the infighting back home was fierce. The Army, the Navy, and the air force all proposed satellite programs. The Air Force effort to promote its Thor rocket was halfhearted, and its Atlas rocket, still being developed, was both risky and expensive. The Army fought hard to become the satellite launcher. It had Wernher von Braun already pushing ahead with a multistage version of the Redstone rocket that would be called the Jupiter, and it had devoted money to a program called Orbiter. But Jupiter was on track to becoming a true ICBM, though its initial Jupiter-C version was smaller.

Its rationale was to subject nose cones—which in a strategic situation would carry atomic or hydrogen warheads—to the fiery heat of coming down through Earth's atmosphere. Letting von Braun indulge his long-held passion for space might slow down Jupiter's development, though how that could be is puzzling. And no matter such thoughts, the Army's rocket program was purely warlike and that was the image Eisenhower wanted to avoid.

The Navy proposed modifications to an upper-atmosphere sounding rocket, with the satellite program sponsored by the Naval Research Laboratory. The NRL had, and still has, solid scientific credentials. Its rocket, to be called Vanguard, was used to launch instruments high into the sky, where they measured winds, radiation, atmospheric density, and more. Vanguard was not a war rocket. The Pentagon's Don Quarles, who had ignored warnings three years earlier that the Russians could humiliate the United States by being first into space, convened a committee. It picked the Navy.

The 1955 decision did not sit well with von Braun, but it was a done deal and was announced in late July by Jim Hagerty. A year later, while the IGY scientists met in Barcelona, von Braun, now technical director for the newly formed Army Ballistic Missile Agency, was preparing for a critical Jupiter test flight. His boss, Major General John B. Medaris, had already gone back to the Pentagon with a new proposal to launch a satellite in early 1957. He was rebuffed.

Jupiter-C was a hybrid rocket. It used clusters of solid-fuel Sergeant rockets developed by the Jet Propulsion Laboratory in Pasadena, California, in its upper stage to get a final acceleration kick. While von Braun readied his Jupiter-C for launch from the Eastern Test Range at Cape Canaveral, Florida, John Simpson and Dr. Herbert Friedman of the Naval Research Lab went to dinner in Barcelona with Russian physicist Leo Sedov. The Russian contingent had been strangely silent during the formal presentations. Simpson speculated later that they were intimidated by the slick American discussion of its Vanguard satellite program.

Whatever Sedov may have felt during the day, he turned voluble that night. A satellite program had received formal approval from the Soviet government, he told Simpson and Friedman. It already was well along in development. The satellite would carry a variety of scientific instruments, he said, and it would be launched in the fourth quarter of 1957. Simpson and Friedman took that to mean October 1957, only a few months after the IGY began on July 1. Sedov even gave them the satellite's radio frequency. It would operate in the fifteen-meter band, virtually next to the time signals broadcast by the American shortwave station WWV. The frequency was selected, Sedov said, because ham radio operators and shortwave enthusiasts the world over used WWV as their master clock. Its frequency was known by heart to all of them, and by moving the dial only a fraction to the right, they would find the Soviet satellite signals.

The American scientists, senior and respected, carried the story home. Simpson told the Central Intelligence Agency. "In my case, there was no time lost," he remembered years later. "You always had an agent arriving [as soon as you got home] saying, 'Well, what's the latest?' " Friedman reported the Sedov conversation to the Naval Research Laboratory, where the Vanguard program was rushing ahead. Neither CIA nor NRL followed up. "It never got to the top," Simpson recalled. Maybe the intelligence services thought it was too incredible. Maybe they doubted the source. Either way, the impact of knowing the Russian plans more than a year in advance was zero. Almost nobody listened.

One of the few who did was Washington columnist Drew Pearson. Word of the Russian plan leaked to him, and for nearly a year, he used his *Washington Merry-Go-Round* column to warn that 1957 would be Russia's year in space. But his columns were dismissed as the rantings of a muckraker.

Inside the Soviet Union, there were plenty of listeners. The coming conquest of space was mentioned now and then in the newspaper *Pravda* and in Tass news service stories. The attention was routine and didn't cause much reaction. It was just a given. The publication that should have been noticed by westerners, particularly Western intelligence services, was given no heed. The magazine was *Radio,* a Russian periodical for amateur radio enthusiasts. It had some circulation outside the Soviet Union, and within, it was read by almost everyone who owned a shortwave receiver. *Radio* gave its readers the same information that Sedov had given to Simpson and Friedman. When the Russian satellite was launched, shortwave radio sets around the world need only be tuned to WWV, then shifted just to the right to hear a signal from outer space.

Less than a week after the Simpson-Friedman-Sedov dinner, Wernher von Braun was in the blockhouse at Cape Canaveral, his Jupiter-C rocket on the pad. He had become an American citizen more than a year earlier and was dedicated to making the United States preeminent in rocket technology. But now he was a frustrated and angry man. His calculations showed that Jupiter's fourth stage could reach escape velocity, about 17,500 miles per hour. At that speed, instead of arcing over and slamming its nose cone back into the atmosphere, it could break the bonds of Earth's gravity and slip into orbit.

Whether von Braun intended it to happen or not may never be known. What is known is that someone in the Pentagon suspected, or had been tipped off, that the world's first satellite might be a military nose cone utterly lacking in scientific instrumentation. It was just possible that the Army could violate orders, put its nose cone into orbit, then apologize profusely for the mistake. So an order flowed down to General Medaris and he complied. Von Braun's Jupiter-C lifted away from its pad on September 20, 1956, at almost the same time that Simpson and Friedman were bringing home the news spilled by Leonid Sedov.

This Jupiter flew more than 3,300 miles into the South Atlantic. At apogee, it was 682 miles high, well out into space. But its speed was only about 16,000 miles per hour. Its third-stage rocket engine burned out, it fell into the classic ballistic parabola and reentered the atmosphere. As a test of nose cone technology, and of America's ability to launch a precursor to a true ICBM, the mission was a great success. The Pentagon dutifully announced the Jupiter-C results, but Wernher von Braun had a bitter taste in his mouth as he accepted congratulations.

What the Pentagon did not announce was that the rocket's fourth stage was inert. Following his orders like a good Army officer, General Medaris had sent technicians to the pad, where they disabled the Sergeant rockets and filled the stage with sand. There would be no accidental satellite put into orbit atop an American war rocket. Nor would the U.S. reputation be "tarnished" by allowing such a stunt. Only a handful of men in the Pentagon had the authority to cause a "don't orbit" order to reach Medaris. One of them was Don Quarles. The word often used to describe him was "cautious."

Whether Quarles was behind the order or not, being first into space, in September 1956, was just not important to the men who controlled American technology policy. America owned the future, but America's bureaucrats were blind. So with the conquest of space in their hands, they turned away and shrank from winning the race they disavowed that they were running.

When the Chief Designer heard the news of von Braun's test, he knew his time was running short. Sergei Korolev believed that von Braun's Jupiter rocket *was* a satellite mission and that it had failed. The U.S. announcement about nose cones and reentry was simply a cover-up. His own satellite program was in deep trouble. Some of the fault was Andrei Sakharov's. When Sakharov, the "father" of the Soviet hydrogen bomb, couldn't deliver reliable estimates of how much—or how little—his hydrogen bomb would weigh, Korolev went ahead and designed the R-7 rocket for the worst case. It would be, at the time, the biggest rocket in the world.

Sakharov eventually delivered his hydrogen warhead, and it was smaller than Korolev imagined. But by then, his R-7 rocket design was locked in and it was big. The satellite being designed for it, however, took full advantage of its payload limits. It was called Object D, and unlike the 22-pounder being planned by the United States, it would weigh more than 3,300 pounds. It would carry nearly 700 pounds of instruments, and the scientific measurements it sent back to Earth would boggle the mind. It would have cameras to photograph Earth and a device to measure the planet's magnetic field. It would take readings of radiation in space, of solar particles, and more. Object D's discoveries would put the rest of the world's scientists in their proper place. Second place.

The plan quickly went awry. Soviet industry couldn't deliver parts on time. Instruments didn't work properly or their operation interfered with other instruments. The Chief Designer had dictatorial control over his R-7 development. He had less control over the satellite. The rocket stayed close to schedule. The satellite fell further and further behind. Sergei Korolev was a realist and he had tunnel vision. He knew that being first into space was the only goal that counted. If Object D with all its scientific instruments couldn't be first, then it was time to change the plan. Korolev thought about reality.

Nearly a year had passed since von Braun's Jupiter-C flight. Inexplicably to Korolev, who continued to believe that it had been a failed orbital mission, the United States had not followed up with "another" satellite attempt. Indeed, von Braun and his rocket had launched two more Jupiters on nose cone tests that were considered partly successful because the rocket worked but the payloads were not recovered and a third, in August, that recovered a nose cone containing U.S. mail. But most attention in the United States was focused on the Navy's Vanguard program. And there had been just two experimental Vanguard flights, one in December 1956, the other in May 1957. Both were successful in testing the rocket, but neither had orbital capability.

Now it was September 1957. The International Geophysical Year had begun on July 1. Object D was hopelessly behind schedule. Surely the Americans would grab victory any day now by fulfilling their promise to orbit a satellite. Despite the fact that, after five failures, he had finally proved his R-7 rocket in August by launching the world's first true ICBM on a 4,200-mile flight to the Kamchatka Peninsula, Korolev's real goal remained out of reach.

He decided to act.

In a matter of weeks, he rammed through approval for a backup satellite that could be launched before Object D. He called it what it was, *Prostreishiy Sputnik*, "Simple Satellite." PS-1 was as simple as they come. The satellite was conceived at Korolev's bureau in the Kaliningrad section of Moscow. It was an industrial area, filled with heavy industry and coated with decades of black soot from smokestacks. Korolev's domain was behind high fortress walls, hiding a complex of hangars, laboratories, manufacturing and fabrication factories, and administration buildings.

What he didn't have for his people in the way of comforts, he also denied to himself. Korolev's office was small and blank. There was a small desk with a chair, and a single old-fashioned black telephone. He had no chairs for visitors. Those who were summoned to Korolev's office stood and listened, talking little unless asked to report. PS-1's plans and diagrams came out of this gray place. It was a 22.5-inch ball of polished aluminum. It weighed 184 pounds and carried no scientific instruments. None. It had a radio that would transmit on the fifteen-meter shortwave band, just next to WWV's time signal. It would also transmit on a second frequency that would let scientists learn something about how radio propagated through the atmosphere. The only other items on board were batteries.

Korolev personally oversaw the satellite's fabrication. The design had been settled for months, and radio tests had been completed in the Moscow bureau. Now two of the Simple Satellites were built in a month at the Baikonur rocket base in remote Kazakhstan. The Chief Designer was a tyrant, driven by the belief that Wernher von Braun would suddenly emerge to put an American satellite into orbit.

Metalworkers stamped out the half-spheres under Korolev's constant attention, with the master craftsmen taking charge of finishing the edges where the halves would be mated and of polishing the outer surface to a brilliant and flawless sheen. Korolev's tyranny even extended to a mock-up satellite used to test separating the satellite from its rockets. In a quote that became famous inside the Soviet program, he berated an engineer because the sphere's shiny surface was not perfect. "This ball will be exhibited in museums!" the Chief Designer snapped. And so it was.

With another S-7 rocket nearing completion, a date was set for launching PS-1. It was October 6, 1957. But Korolev's obsession with being first intruded again. He received an agenda for the Eighth Congress of the International Astronautical Federation. It was at this gathering in 1955, in Copenhagen, that Leonid Sedov announced plans for a Russian satellite. Now the Congress was convening in Barcelona. One paper caught Sergei

Korolev's eye and sent fear into his heart. It was an American presentation and it would be given on October 6: "Satellite Over the Planet."

Korolev saw only one reason for the paper's title. An American satellite would be in orbit when it was delivered. He must do something, and fast! A hurried check of his R-7's status showed that it could be rushed to the pad on October 3 and be launched a day later. He ordered it to happen. Russian rockets were not moved to their pads until the last moment. They arrived aboard trainlike carriers and were lifted to the vertical. Korolev walked ahead of his rocket—*his space rocket!* This taciturn and difficult man, this Chief Designer, led the way to the distant pad.

He had to know that man's first steps into space were literally his own.

An artillery lieutenant named Boris Chekunov pushed the button. The engines ignited at 10:28:34 p.m. Moscow time, October 4, 1957, a Friday. The Chief Designer was one of only two men in the Tyuratam control room who sat. The other was Korolev's deputy, Nikolai Pilyugin.

Nikita Khrushchev was in Kiev, on his way home to Moscow from a vacation at the Black Sea. His son Sergei, only twenty, was with him in a room with Ukrainian party leaders. The Ukrainians were arguing for additional funding from Moscow to support their local programs. Andrei Gromyko, the Soviet Foreign Minister, was in Washington. He had been invited, *commanded,* to appear the following day, a Saturday, at the home of Secretary of State John Foster Dulles. In the wake of its hydrogen bomb and ICBM successes, the Soviet Union was rattling sabers along the Turkish border. Dulles would tell Gromyko that any attack on Turkey would lead to war with the United States. But first, Gromyko would attend a party on the night of October 4 at the Soviet embassy.

Things went wrong with the R-7 rocket almost immediately. Several engines ignited late. A fuel-flow system failed and a turbine shut down the main engine one second too soon. But it continued on course. Khrushchev drank vodka and attempted to convince the Ukrainians of the country's financial realities. Caterers brought food and drink to the Soviet embassy in Washington and began to set up for the Friday night party.

The R-7 rocket and its Simple Satellite payload flew out of radio range. Four thousand miles away, radio operators at the Kamchatka station sat with earphones on, their shortwave sets tuned to the proper frequencies. The sound came: *Beep. Beep. Beep.*

At Tyuratam, they cheered until the Chief Designer held up his hand. The celebration would wait until PS-1 came around on its first orbit and they heard the *beep-beep-beep* for themselves. It was a long ninety minutes. And then it was there.

Korolev picked up the phone to call Nikita Khrushchev. In Kiev, an aide stuck his head into the room where a rancorous budgetary argument was raging. "That call you've been waiting for," the aide said. The premier of the Soviet Union left the room but was back in minutes. He was beaming.

"The Soviet Union," he announced, "has just put the world's first artificial satellite into space around Earth."

What it meant was beyond Khrushchev's understanding at the moment. He knew only that it was a fine scientific achievement and that Sergei Korolev's war rocket had worked again. The Ukrainians understood even less. Opening a new power plant and christening a new airliner were also important technological events, and that this particular event could have some transcendant value occurred to no one in the room. There was polite applause. But the premier was in town only for a brief time, and the issues to be discussed were vital. Having Khrushchev's ear was a rare opportunity, and they wanted to get back down to business.

"That's nice," the Ukrainians said of this thing somewhere in outer space. "Now what about our budget?" But it was late and Khrushchev was in no mood for more pleadings from regional bureaucrats. He took another vodka and went to bed.

Shortly after, the Soviet news agency Tass moved a short and extraordinarily bland announcement on its wires. The news went out to the world, and *Pravda* carried it verbatim at the bottom of its front page the next morning:

For several years scientific research and experimental design work have been conducted in the Soviet Union on the creation of artificial satellites. As has already been reported in the press, the first launching of the satellites in the USSR was planned for realization in accordance with the scientific research program of the International Geophysical Year.

As a result of very intensive work by scientific research institutes and design bureaus, the first artificial satellite in the world has been created. On October 4, 1957, this first satellite was successfully launched in the USSR.

According to the preliminary data, the carrier rocket has imparted to the satellite the required orbital velocity of about 8,000 meters per second. At the present time the satellite is describing elliptical trajectories around the Earth, and its flight can be observed in the rays of the rising and setting Sun with aid of very simple optical instruments (binoculars, telescopes, etc.).

The story did not name PS-1, referring to it only by the Russian word for satellite: *Sputnik.* So it became.

News reached the United States shortly after the party began at the Soviet embassy. The ambassador and his staff were caught by surprise, but no more so than American officials and journalists already sipping wine and vodka and eating crackers spread with caviar. Reporters rushed to get reaction. They got plenty. Saturday morning's *New York Times* was not bland. A black three-line, eight-column banner headline spread across the front page:

SOVIET FIRES EARTH SATELLITE INTO SPACE;
IT IS CIRCLING THE GLOBE AT 18,000 M.P.H.;
SPHERE TRACKED IN FOUR CROSSINGS OVER U.S.

Four front-page stories gave details of the Russian achievement, along with jumps and more stories and diagrams on inside pages.

A few days later, Don Quarles, now a deputy secretary of defense, told President Eisenhower in a National Security Council meeting that Sputnik was a good thing for the United States. Said the minutes of the meeting: "[Quarles said] that the Russians have in fact done us a good turn, unintentionally, in establishing the concept of freedom of international space—this seems to be generally accepted as orbital space, in which the missile is making an inoffensive passage." This was equivocation of the highest order. But no one in the Eisenhower administration was yet willing to face the reality of what had happened. To see the truth, they had only to pick up any newspaper.

An age of naïveté in the United States had ended. A space race would take its place. Historian Walter A. McDougall coined a name for the transition between the two.

He called it a media riot.

2

Nikita Khrushchev knew a good thing when he saw it. The second-day stories in *Pravda* and every other Soviet publication were blockbusters. In the United States, it seemed that everyone from schoolchildren to newspaper reporters to politicians was bemoaning national failure. The Soviet Union had a major propaganda victory on its hands—the biggest victory it would ever have from the moment of its inception by Lenin and the proletariat in 1917 to its demise at the hands of Ronald Reagan in 1989—and it knew how to play that game too well.

The immediate question on Khrushchev's mind was what to do next. He didn't want Sputnik to be a momentary triumph. With the fortieth anniversary of the revolution and the founding of the Soviet Union to be celebrated in less than a month, he wanted another victory in space. Sergei Korolev, at the Kremlin to receive Khrushchev's personal thanks, barely hesitated before promising that he would deliver another Communist victory in time for the big party. Sputnik 2 roared into space on November 3, 1957, and the world gasped anew at the 1,120-pound behemoth. This was no simple satellite, though it was another of the Chief Designer's one-month wonders. A passenger was on board Earth's second artificial moon. Her name was Laika. The mongrel dog would survive seven days before her oxygen supply ran out, providing the first proof that life could be sustained in zero gravity. There was another message here too. The Russians clearly were looking ahead to putting a man into space.

Reaction in the United States was swift and riotous. Congress, the press, and the man in the street, down to the shortest street in the smallest town in the country, demanded to know how this had happened, how the bumbling, backward Russians had managed to humiliate America twice in thirty days by orbiting satellites that could be seen, on clear nights, from anybody's backyard as they crossed the sky. The phrase "missile gap" began to creep into news stories, into political speeches, and into the American consciousness. Newspaper headlines, radio commentators, and the early television news readers all screamed the same question at the Eisenhower administration: How can this be?

In Moscow, Nikita Khrushchev, chairman of the Communist Party and premier of the Soviet Union, spoke proudly to the Supreme Soviet. "Our sputniks are circling the world," he said, pointing upward. He boasted that Soviet technology now led the West, and in a phrase that seemed to be a throwaway to American analysts, told the leadership of the Soviet Union that war was not the way to victory. "Wars are not needed," he thundered. The socialist system was proving its superiority in other ways. He was serious and he believed every word he spoke that day. His message was lost. Mutual mistrust and profound misunderstanding between America and Russia ruled.

A few days after Sputnik 1, Don Quarles told the president that America could have easily beaten the Russians to orbit. At the moment that Eisenhower mulled this admission, Wernher von Braun was setting up a howl of his own in Huntsville, Alabama, begging to use the Army's Jupiter-C missile to launch an American satellite instead of waiting for the Navy's Vanguard program to do the job.

The military-rocket-versus-peaceful-rocket question continued to bother Eisenhower. Congress would soon know what Quarles was telling him, but he wanted to maintain his policy of using Vanguard for the peaceful International Geophysical Year missions. Scientists were to be kept away from the secrets of military rockets like von Braun's. And he brightened when Quarles pointed out that, whether the Russians intended to or not, Sputnik 1 settled the question of orbital overflights once and for all. Outer space, like the high seas, was open to free passage by all nations. That was key to all of Eisenhower's thinking about space. Science and the IGY were nice, but the true value of the high frontier was for reconnaissance. Still, the media riot that was just beginning made national prestige the focus of every argument. Something had to be done.

Von Braun got a pale green light five days after Sputnik 2 to begin preparations for launch, but the Navy maintained its place at the front of the line.
Von Braun's Jupiter team was strictly a backup. Eisenhower's press secretary,
Jim Hagerty, had told reporters after Sputnik 1 that a Vanguard satellite
launch would be attempted in December. The pressure on the Navy was
intense. A next Vanguard test flight was scheduled for early in 1958, but it
was rushed ahead, topped with a tiny 3.25-pound, 6.4-inch spherical satellite, and moved to the pad at Cape Canaveral, Florida. Even then, the
Vanguard launch team was in a bind. They considered the launch to be a
rocket test, not a full-fledged satellite mission. But Hagerty's press-priming
in October made that subtle distinction irrelevant to the American public.
The Vanguard launch had morphed from a previously scheduled test flight
into Uncle Sam's answer to the Russian bear's challenge.

It was one of the few times in the early days of rocket launches that the
government gave more than twenty-four hours' advance notice to the press.
This time it barely mattered. Despite every attempt at security, some heavy-
handed, some subtle, little happened at Canaveral without reporters in
nearby Cocoa Beach learning about it. Bartenders and waitresses passed
along overheard conversations, motel clerks noted the influx of military and
corporate customers, and the traffic along Highway A1A thickened whenever a flight of any kind was imminent. By late November, with two Sputniks beeping merrily high overhead, the area around Cocoa Beach was
getting crowded.

So when the Naval Research Laboratory formally announced that it
would launch a Vanguard satellite on December 4, 1957, the press was
already in place or on its way. Television and news cameras set up at the
official press site, on the beach, and even on the roofs of Cocoa Beach motels
and condominiums. *Life* magazine sent veteran photographer Ralph Morse.
Harry Reasoner was there for CBS. The *New York Times* had a man on scene
and so did the *Los Angeles Times*. The Associated Press and United Press
International had their Teletypes on-line from Cocoa Beach, with editors at
newspapers in every city in the land, and most cities overseas, standing by
for the story that would be one of those rare ones carrying the *Flash* designa-
tion. Even columnist Dorothy Kilgallen came down for the event. None of
them understood how difficult it is to launch a rocket.

But to the Vanguard people, this *was* rocket science and it wasn't easy.
Deputy Vanguard director J. Paul Walsh tried to put the test flight into

perspective for reporters. "We'll be pleased if it goes into orbit," he said in a last-ditch attempt to defuse the public frenzy. "We'll not be despondent if it does not." The subtle distinction between a test flight and an all-out effort to join the Russians in space was lost on the reporters, and so on the rest of the country too. With the eyes of the world on Cape Canaveral, the countdown finally began, seven and a half hours late, at 4:30 a.m. on December 4. It didn't go smoothly. One delay after another pushed the launch time later and later into the day. Finally at 10:30 p.m., with a critical valve frozen shut and high-altitude winds gaining strength, the mission was scrubbed. Reporters filed their stories, radio and television people signed off, and the bars at Cocoa Beach did a glum but profitable business into the wee hours.

They started over again the next afternoon, a Thursday, hoping to see Vanguard off the pad at 8 a.m. Friday. There were more problems, more holds in the countdown, but nothing bad enough to force another cancellation. Eight o'clock came and went, and it was 10:30 a.m., an hour before liftoff, before Vanguard's gantry began to slowly move back. A warning horn sounded at T minus thirty minutes, its mournful blare echoing across Cape Canaveral's palmetto landscape. The few technicians left near the pad raced away to safety. Vanguard stood alone. The modern control centers that would become so familiar had not yet been dreamed of. A simple blockhouse, with heavy doors and thick windows, was filled with key Vanguard people. They surrounded some simple consoles that controlled the countdown. The firing button was just a toggle switch, not a button at all. With one second left in the count, a young engineer named Paul Karpiscak flipped the switch.

The next few seconds made history. Ugly history. Vanguard spouted flame from its tail, shivered in high-tech desire, lifted a few feet into the air, and dropped back. Before the sound of its engines could reach spectators, they saw its long body crumple, fold, and explode in a brilliant ball of orange fire and black smoke. The little satellite was thrown free, its tiny palm-sized transmitter activated by the release, and it lay on the ground nearby beeping a signal that no one outside of the blockhouse could hear. An investigation into the failure was "indeterminate." It was caused either by low pressure in a fuel tank or by a fuel line fitting loosened by technicians who stood on it.

To the rest of the world, it didn't matter. This was a public disaster of the worst sort. *Pravda* chortled a snide "Publicity and Reality," while the *Lon-*

don Daily Herald ran a huge banner headline over a photo of the Vanguard's final fiery seconds: "OH, WHAT A FLOPNIK!" In New York City, Russian delegates at the United Nations suggested that the Americans take advantage of a Soviet program offering technical assistance to backward nations.

The American press was scarcely less kind, and Lyndon Johnson led the baying hounds of Congress in blasting the Eisenhower administration for the failure. The Vanguard debacle, Johnson said for the record, was "most humiliating." Senator Hubert H. Humphrey of Minnesota agreed. In a biting statement, he said that he was "saddened and humiliated by the cheap and gaudy manner in which the Administration has gone about the business of trying a last minute rush to launch a satellite." Most Americans agreed.

Two did not. In a polite exchange with the president of the National Academy of Sciences, President Eisenhower pointed out that Americans didn't despair when a British scientist discovered penicillin first. Ike still hadn't realized that the space race he wasn't running was already the worst disaster of his presidency. He had two more years in office for that bitter fact to sink in. Vice President Richard Nixon, on his way to becoming a near-fanatical supporter of space exploration, was more realistic. He wrote to the Vanguard program's youthful director, John P. Hagen, commiserating that Vanguard was "catching it from all sides," and promising that the bad times were temporary. ". . . I want you to know that I, for one," Nixon wrote, "feel you should have every support."

But Hagen's team had blown its chance, though it was a chance they weren't quite ready to handle and didn't really want. They would try again. But the next time Vanguard sat on a Cape Canaveral launch pad, another rocket and another satellite would be standing by just in case.

"Our German" was about to take the field.

Wernher von Braun was in midlife, forty-five years old, a naturalized American citizen, resident with other Germans and former Germans in the small town of Huntsville, Alabama, and about to embark on the crowning adventure of his life. He would be one of America's designated runners in the space race. His genius, both as a rocket engineer and a manager, had given Adolf Hitler's war machine a series of terror weapons—ballistic missiles—that culminated with the V-2. In the last months of the war, V-2s rained down on cities in England, France, and Belgium, bringing fear and loathing, but doing insignificant damage and nothing to turn the tide of battle. World

War II was lost and von Braun knew it. In 1945, he and much of his team, along with major stores of rocket hardware, surrendered to American troops.

Now it was 1958. The V-2, with more than a decade of improvements and modifications, became the Army's Redstone rocket. A Redstone, with three additional rocket stages stacked on top, became the Jupiter-C. Only by filling the top stage with sand instead of solid rocket fuel had the Pentagon, carefully keeping this secret from the press, stopped von Braun from putting a Jupiter nose cone into orbit in 1956. The United States of America was not going to make that mistake again.

Hours after Sputnik 1 reached orbit, von Braun vehemently predicted that the Vanguard program would not be able to orbit a satellite in the near future. In an Oval Office meeting, President Eisenhower shook his head and said that he was tired of "von Braun's pronouncements from Huntsville." Ike's antipathy didn't stop the chief of the Army Ballistic Missile Agency, Major General John B. Medaris, from allowing von Braun to begin preparations anyway. When the official word came down a month later ordering "our German" to launch a satellite as soon as possible, von Braun was well into the process. Eisenhower heard that Medaris had turned von Braun loose early, and fumed that a court-martial was in order. But he settled down quickly and did nothing that would keep the Army from its new assignment in space.

The Army had a rocket. It needed a satellite. Von Braun got it from the Jet Propulsion Laboratory in Pasadena, California, where Dr. William H. Pickering ruled and had people working on a satellite to be called Explorer. It weighed just shy of thirty-one pounds, including eighteen pounds of instruments. Explorer was hardly in the weight class of either Sputnik. But it was what they had. Pickering was well-known in broad science circles. The man who designed and delivered Explorer's key scientific package was more of a specialist. He was Dr. James Van Allen of Iowa State University. The singular instrument in the Van Allen gear was a cosmic ray detector, a sort of Geiger counter, designed to measure radiation above Earth's atmosphere. A radio transmitter to send data home was part of the package.

But still, von Braun and his Army team were the backup. The Vanguard program had a new rocket on the pad, a new satellite in its nose cone, and high hopes of a launch in the fourth week of January 1958. What the Russians knew is an unanswered question. What the free people of the world knew is a matter of record. They knew nothing about it. Air Force Major General Donald Yates, who commanded the Missile Test Center, took enormous steps to tighten security. Beginning a few days before Christmas 1957, he banned all cameras, television and still cameras alike, from the beaches north and south of the Cape Canaveral launch area. He banned binoculars too. Newsmen assigned to Cocoa Beach didn't raise a fuss. The catchphrase that guided all sorts of good and bad reporting after Watergate had not been created. *"The public's right to know"* didn't exist.

In mid-January, Yates and his security staff ran a bluff on reporters. "The country's prestige got burned around the world by the Vanguard failure," newsmen were told at a hastily called, off-the-record briefing. If reporters didn't cooperate with the tightened security, the Air Force would just make security even worse. Reporters went away muttering, but they didn't write stories about it.

The rooftops of Cocoa Beach motels and office buildings were still private property. Reporters and photographers could set up vigils, and the Air Force would be powerless to stop them. And they all knew that the bright balls atop a pair of ninety-foot towers at the Cape were signals of an imminent launch. It was cold and windy and often wet atop those buildings, but an alert observer would see the balls drop and have time to alert the rest of the press corps that something was about to happen. The next day, reporters were called back for another session. "Here's the deal," an Air Force officer said. "We'll give you advance notice of rocket firings if you agree not to write a word until the launch is successful."

The influx of Vanguard people into Cocoa Beach was a big clue that something was coming, and soon. If reporters could avoid taking their turn on rooftop duty, particularly during the evening drinking hours and the cold overnight tour from midnight to dawn, by accepting the Air Force terms, maybe it was a good thing to do. The deal was struck. Home offices were notified and distant editors agreed to the plan. It was time to wait. Within a few days of the formal agreement to hide rocket launch activities from the public, the Air Force lived up to its part by telling reporters that a Vanguard countdown was about to begin. Nobody wrote the story or blurted it out on radio or television. There were problems and the countdown was scrubbed. Nobody wrote about that either.

A new launch date was set. The countdown began again and was stopped again. And again. And again. Sometimes the cause was bad weather. Sometimes it was a malfunction of a part on the rocket, or an instrument that gave readings beyond the limits. Then near the end of January, with still not a word leaking to the American public, the countdown got all the way down to T minus 14 seconds before it was stopped. Reporters had open lines to their offices, television and radio reporters were ready to go on the air when the Vanguard satellite reached orbit. They wrote nothing and they said nothing when the mission was scrubbed.

This time the problem was bad. The rocket's second-stage engine would have to be replaced. It would be at least three weeks before Vanguard could try again. Too many people, civilians as well as reporters, knew what had happened. The air force put out a short statement for the record saying that a Vanguard rocket had encountered difficulties on the pad and that its launch would be rescheduled for an unspecified date. As reporters were thinking about going home for a while, the air force called again. The Army and Wernher von Braun's team were moving a Jupiter-C rocket to a launch pad.

The Army's turn was set for January 29, 1958.

Wernher von Braun wasn't there. At General Medaris' order, both he and Bill Pickering were in a communications room at the Pentagon, politicking and figuratively holding the hand of Army Secretary Wilber Brucker. In the Soviet Union, the Chief Designer was at the site when his rockets took off for space. In the United States, schmoozing the brass was more important and von Braun was livid. But he followed orders. This man who had met Hitler (and been briefly jailed by him), bombed London, been lionized in American magazines, and dined at the White House knew when to put his anger aside and play an assigned role.

In von Braun's place was one of his key deputies, Kurt Debus, whose experience went back to Peenemünde and who had fired more experimental rockets than anyone in the world. At dawn on January 29, Debus looked at the weather report, saw jet stream winds at 170 mph, and canceled the countdown. Reporters spent the day drinking and thinking about the stories they couldn't write. Von Braun went to his hotel in Washington and fretted. The next day, winds were 205 mph at forty thousand feet. Debus canceled again.

On the third day, winds aloft were at 157 mph and dropping. Jupiter-C was a science hybrid of a military missile. The C version wasn't quite as tough as a fighting Jupiter, which was designed to be launched in almost any weather and wind speed. Rocketeers couldn't pick the conditions when a counterattack against Russia might be required. But it was tough enough. Debus said the magic word: "Go."

It took almost all day plus eight minutes, and then another hour and a half, to make history. While Wernher von Braun all but held his breath in that Pentagon communications room, and Kurt Debus shielded his eyes and looked skyward, a few dozen reporters and broadcasters scribbled notes and uttered recorded descriptions into microphones, Jupiter-C lifted off, soared away on its tail of fire, and disappeared into the black Atlantic sky. The rocket left the pad at 10:55 p.m. Its pinpoint of light was gone to the naked eye, then to binoculars, a few minutes later. The trip from Florida to orbital space was to take eight minutes. If everything worked, Explorer 1 would be in orbit at about 11:03 p.m. eastern standard time on January 31, 1958.

How distant and primitive that day seems in the light of what followed. There was no worldwide tracking network. Reports came in that an Explorer-like signal had been received in the British West Indies. Another fifteen minutes passed before word came from Ghana, in West Africa, that signals were being heard. But there was no confirmation of an orbit. No data relay ships were stationed in the South Atlantic, the Indian Ocean, or the Pacific. Certainly there were no communications or surveillance satellites at geosynchronous orbit, their sophisticated instruments watching Explorer every inch of the way. It was on its own, and so were the people on the ground. They could only wait until 12:41 a.m., when an orbiting Explorer would come into view of a tracking station in San Diego and its radio would either be heard or not.

There was jubilation at the launch news, apparently perfect. Now came the wait. President Eisenhower played bridge with friends at his winter White House in Augusta, Georgia. Kurt Debus paced the blockhouse at Cape Canaveral. Reporters smoked, napped, gossiped in the rudimentary pressroom in a Cocoa Beach motel. Wernher von Braun and Bill Pickering chatted with Secretary Brucker and a handful of favored generals and colonels in the Pentagon. Von Braun said later that he thought he could hear his own heart beat over the forcedly casual conversation.

When the big wall clock ticked close to the time, Pickering picked up a phone. San Diego was on the other end. At 12:41 p.m., there was silence. At 12:43, San Diego reported more silence. At 12:49, there was the silence of despair. Brucker begged von Braun for an explanation, and the officers in the room edged away from our German. Von Braun could only look stoically at Pickering, phone still pressed to his ear, and wait in silence. Then Pickering's slumped shoulders straightened. His face went from funereal to radiant. Explorer was coming in loud and clear. San Diego had the satellite locked in.

It was 12:49 p.m. The Jupiter-C rocket had delivered a bit extra, and Explorer was in a higher orbit than planned. It took an extra eight minutes to make the circuit from Cape Canaveral to San Diego. Von Braun took a deep breath and began shaking hands. In Augusta, President Eisenhower made the announcement himself. If he was excited, it didn't show. "The United States has successfully placed a scientific Earth satellite in orbit around the Earth," he said into a live microphone. "This is part of our participation in the International Geophysical Year." There was more, the customary congratulations to the team and so forth. Nobody much cared about that. Euphoria coast to coast was the national mood.

The space race had begun for real.

A few weeks later Wernher von Braun delivered a speech to the National Military Industrial Conference in Chicago. At the same time, Jim Van Allen's Geiger counter aboard Explorer 1 had recorded an intense belt of radiation girdling the Earth. Years of study, and more satellite missions, would show that a series of these Van Allen Belts were up there and that they act as a sort of trap, keeping much harmful radiation from getting down to where we live.

So Explorer 1 met Eisenhower's goal of being a science mission. But what the nation really cared about was getting a satellite, any satellite, into orbit. The man who had done it was the same man who had been simultaneously praised and excoriated five years earlier in *Time* magazine for being a visionary with his eyes on space. Eisenhower, who'd made no secret that he didn't much like von Braun, shook his hand in the White House and presented him with the Distinguished Federal Civilian Service Award. And the fledgling National Rocket Club awarded him its first Dr. Robert H. Goddard Memorial Trophy. *Time* put Wernher von Braun on its cover and the rocketeer rose to speak in Chicago. Our German was the man of the hour.

He was 100 percent American when he spoke in Chicago on February 17, 1958. "Our country has faced agonizing tests more than once during its relatively short history," von Braun said with a new degree of fire and passion. "It emerged each time from the crucible not without scars but with greater confidence and richer maturity . . ." The space race was a turning point, von Braun said, and of historic proportions. There was no way to go back.

"The reaction to these events has been profound. They triggered a period of self-appraisal rarely equaled in modern times. Overnight it became popular to question the bulwarks of our society: our public educational system, our industrial strength, international policy, defense strategy and forces, the capabilities of our science and technology. Even the moral fiber of our people came under searching examination. Since the evident threat was to our security, the initial preoccupation concerned modern weapons systems and means of defending against them. The Damoclean sword menacing free people consisted of a monstrous destructive force inherent in automatic delivery systems, capable of transporting thermonuclear warheads thousands of miles, in *any* weather, across all geographic and political barriers, at velocities of such magnitude as to imply total destruction without advance warning."

In a single paragraph, requiring no more than a few seconds to speak, von Braun made it clear that rocket science had military science at its heart. Eisenhower's fervent desire to publicly separate rocketry from war was, at least for the opening laps of the space race, a pipe dream. But von Braun then broadened his vision, delivering an eloquent and accurate portrayal of the race yet to come.

"The Soviet challenge is by no means restricted to military technology. It goes far beyond the realms of politics and armies. No longer is the task of coping with the Red menace the exclusive responsibility of generals and statesmen. The acid test involves every facet of our civilization, every part of our society: religion, economics, politics, science, technology, industry and education. Free men everywhere have been caught up in this grim competition . . . What we are about to discover is whether a nation, who has rated its home run sluggers and its fullbacks above scientists and philosophers, can meet the total competition of aggressive communism, and still preserve its way of life."

With searing words delivered from the mind and the heart, von Braun accurately sketched the future's outlines on a canvas that would be painted and filled in by himself and others over the course of the space race, and beyond.

"It must be understood also that the Soviets have grasped the significance of man's imminent conquest of space and have proceeded well along the road in that direction. A current estimate of the situation would include these possibilities:

"First, the Russians probably used a multistage rocket to launch their satellites which was originally designed to carry a thermonuclear warhead over intercontinental range.

"Second, the same rocket configuration, with minor modifications, can place a payload of between 50 and 100 pounds on the moon.

"Third, the rocket can also put a satellite capable of military reconnaissance, equipped with a television playback feature. A few such orbital devices can keep track of the progress of all surface construction projects, ship movements, and air base operations anywhere in the world. Once they achieve this, and I am convinced that is only a very few years off, 'open skies' inspection for purposes of disarmament becomes academic.

"Fourth, the Russians have a sound program designed to solve the question of safe return from orbital flight and related space medical problems, with the purpose of preparing for manned space travel.

"I would recommend that we brace ourselves for other Soviet 'firsts' in the new field of astronautics. We are behind and we cannot catch up in a day or two, since major technological projects necessarily involve lead time. It will require years of concentrated effort to come abreast, and even longer to pull ahead."

He spoke of education as the source for the scientists and engineers needed to confront the world realities, and of the emotional content of the space race, "the will to supremacy. Because this is intangible, because it must come from the hearts and minds of our people, it cannot be legislated, budgeted or evoked by decree." But it could be brought forward by fear, and whether he planned to or not, von Braun on that day was doing as much to promote national fear as he was to inspire a national determination.

But von Braun knew, too, that Americans need more than fear to be at their best. They want a mission, a higher goal than simple self-preservation. In the final few seconds of his talk to the National Military Industrial Conference, he was speaking to all Americans.

"No man can say with assurance what benefits will accrue from our discoveries. With Explorer, we made a modest beginning. We have stepped into a new, high road from which there can be no turning back. As we probe farther into the area beyond our sensible atmosphere, man will learn more about his environment; he will understand better the order and beauty of creation. He may then come to realize that war, as we know it, will avail him nothing but catastrophe. He may grasp the truth that there is something much bigger than his one little world.

"Before the majesty of what he will find out there, he must stand in reverential awe. This, then, is the acid test as man moves into the unknown."

The unknown waited for the next steps in the race.

On a January day in 1937, twenty-four-year-old Bob Gilruth boarded the train in forty-below weather in Duluth, destination Hampton, Virginia. He'd borrowed twenty-six dollars from his father for the ticket. He'd never been to Virginia, but it had to be warmer than Minnesota. Gilruth, with a fresh master's degree in aeronautics from the University of Minnesota, pulled a cloth sack of Bull Durham from his pocket and rolled the first of many cigarettes he smoked in the next fifty-six hours. He changed trains in Minneapolis, Chicago, and Chattanooga before rolling into Hampton bleary-eyed, hoarse, and excited.

There was a place on the edge of town that drew the best of the best in the young field of aeronautics. Langley Field was home to an airport, a small squadron of Army airplanes, and a collection of buildings, most of them wood- or metal-sided, that housed a research organization called the National Advisory Committee on Aeronautics (NACA). The committee itself was a pantheon of aviation legends. Orville Wright was still active. The youngest committee member was Charles Lindbergh. Others included Joe Ames, Harry Guggenheim, and Gus Robins, all pioneers in flight.

Bob Gilruth had a job waiting at NACA. He arrived with impressive credentials for a fellow reporting to his first real job. At Minnesota, he'd been a graduate assistant under Dr. Jean Piccard, the famed balloonist, and Piccard's equally famous and possibly smarter wife, Dr. Jeanette Piccard. He'd spent another term helping to redesign Roscoe Turner's airplane for the 1936 National Air Race. His master's thesis was about a scheme to use propellers mounted at wingtips to increase an airplane's speed and stability. It wasn't a very good idea, but the research he put into it was meticulous.

At Langley Field, the clerks looked at his degree, aeronautical engineering, and assigned him to NACA's Aerodynamics Division. Gilruth was happy with the assignment. He thought that any job was a good job in the Depression and any job in aviation was like being entered into heaven. He'd only been in Aero a day when the boss called him in. Gilruth's experience in stress analysis and airplane design wasn't what Aero needed. The young Minnesotan was sent packing over to the Flight Research Division.

At Flight Research, engineers worked with real airplanes. They were side by side with test pilots. Theory was important, but the key to a job at Flight Research was finding what worked. And fixing what didn't. When a test pilot encountered trouble, the engineers on the ground worked with him, if he was still alive, to make sure it didn't happen again. If a test flight went well, the engineers interrogated the pilot and looked for ways to make it even better the next time up. Bob Gilruth found himself in the real world of experimental aviation, getting a look at the big picture of flight as well as the nit-picking and sometimes bloody details.

At first he had nothing to do. His new boss wasn't expecting him, assigned him to sit at one of the eight desks in the small office, and ignored him. "Being a person who didn't want to look like he didn't have anything to do," Gilruth said, "I went and got a lot of books and started studying the things I thought I was going to need to know. I read all the technical reports that Flight Research had written, and I tried to learn what they did and what their instruments were like. I worked hard doing that thing, and nobody paid much attention to me." So he learned about Flight Research, but at the same time he was bored. That was another lesson he never forgot. In later years, when Bob Gilruth became a power in aviation, and then space, "When some junior engineer was sent over to my jurisdiction, I gave him something to do right away."

After a week of reading, an engineer named Hartley Soulé walked over and dumped some rolls of film on Gilruth's desk. "Here, you aren't doing anything," Soulé said. "Work these up." Soulé had been flying with test pilots, recording instrument readings on film to pin down the flying qualities of various airplanes. In the end, Gilruth's job was to determine what made airplanes fly the best—what kind of controls, what kind of pilot inputs, what sizes and shapes for wings, elevators, ailerons, and flaps, even fuselages. If such data could be quantified, it could be made into a kind of encyclopedia and rule book for aircraft designers. It would eliminate much, though never all, of the trial-and-error style of designing airplanes.

A small fleet of airplanes was flown for the project, everything from airliners to private planes to Army and Navy combat planes. Soulé had the exciting part of the job, riding along on the test flights while Gilruth analyzed his films and helped him write reports. Still, it was a lucky break for Gilruth. This was an important project and he devoted himself to it.

Then he got two more breaks in quick succession. Soulé developed a sinus problem and couldn't fly. Bob Gilruth took his place in the air, quickly developing an intense fascination for flight, high speed, and actual control over an intricate machine. He badgered the test pilots to let him fly some of the routine parts of the mission. He flew a Martin B-10B, the first low-wing monoplane bomber, and a one-of-a-kind Boeing B-15. NACA's future chief test pilot, Mel Gough, took a liking to Bob and showed him things about flying that were beyond most engineers' ken. One day the two of them even crammed themselves into a single-seater and took off, Mel's left foot on one rudder pedal and Bob's right foot on the other. But Gilruth never took the next step to becoming a pilot himself. He just soaked up the feelings of controlling an airplane in a bank or a dive or a climb or a sideslip, building his experiences and intuition. Then he turned the controls back to the test pilots when the real maneuvering began so that he could run the camera and focus his attention on the instruments.

As he secured his position and his reputation, Gilruth indulged himself in another passion, finding tricks to make sailboats go faster. He'd read somewhere as a kid that before Alexander Graham Bell invented the telephone, he had been granted a patent on something called hydrofoils, sleek pontoonlike shapes that would let a boat rise up out of the water and virtually glide over its surface. Gilruth contacted Bell's descendants, got permission to examine Bell's original drawings in the U.S. Patent Office in nearby Washington, and even to play around with modifying and building some hydrofoils of his own.

Boatbuilding became Bob Gilruth's pressure release valve. He designed hydrofoil hulls at night, built them with his new wife in their small living room, and eventually took his design out into Chesapeake Bay. Settled into the water, his boat looked awkward and funny. Other sailors did the unforgivable. They pointed and laughed. When they'd jokingly challenge Gilruth to a race, he'd nod and say, "Sure, let's go."

Then he'd catch the wind in his sail, his little boat would lift up until its hydrofoil edges barely touched the water, and he'd send it skimming across the bay, sometimes two or three times faster than the racing yachts that challenged him. His laughter and shouts of joy carried back to the yachtsmen on the singing salt air. For the rest of his working life, Bob Gilruth did flying machines by day and boats by night. Some of his patents are still used today in high-speed hydrofoils.

Another lucky break came when Soulé was promoted to another division and Gilruth took over, finding himself at age twenty-five on the forefront of flight testing. When no one replaced him and no one told him to stop, he began building and then expanding NACA's flight testing section. The data he, and now the members of his team, developed found their way into the hands of aircraft designers at Lockheed, Boeing, Douglas, North American, and all the other airplane companies that were about to enter a long period of frenzied design, test, and production. Important things were happening in the world as 1938 faded and 1939 took hold.

A war was about to start. It would be a war that fostered a great leap forward in aviation, and a war that settled the future for men like Bob Gilruth and Wernher von Braun and Sergei Korolev, and in truth, for many of the rest of the world's residents too.

Their war careers were decisive. But only Wernher von Braun came out of World War II with an international reputation as a rocket man. Sergei Korolev scooped up a few key Germans and a trainload of minor technicians, moving them to the outskirts of Moscow in 1946. He was the Chief Designer, a man with no name.

The third man in the space race triumvirate was Robert R. Gilruth. Nobody knew him either. He came out of the war as a respected expert in aviation. He and his team at the National Advisory Committee on Aeronautics provided much of the research into designing and refining high-performance airplanes that helped give the Allies air superiority by the middle war years. His reports so impressed the British that they assigned liaison experts full-time to Langley Field in Virginia, where they could benefit from

Gilruth's growing expertise firsthand and relay his findings rapidly back to England.

Of these three—von Braun, Korolev, and Gilruth—one would be the father of American moon rockets. Another would be the father of Sputniks and cosmonauts in orbit. And the third would be the father of America's man-in-space program. Only von Braun would become a household name and he would be the least of the three.

Bob Gilruth was a nobody. But Bob Gilruth was a nobody with a curious mind. The war ended with aviation on the cusp of great things. American B-29 bombers had flown at the edge of the stratosphere. The Germans had flown jet-powered airplanes and bombed London with rockets. The Russians were far behind on both fronts but were determined to catch up. Only the most reactionary scientists still thought that the sound barrier was a real barrier to winged airplanes. Gilruth didn't know it in 1945, but he was beginning to put together the nucleus of a team that would do great things in a future that was not so far, far away. Gilruth's men just needed to learn about things that had yet to be discovered and to invent technologies that had barely been dreamed of. One fellow who worked for Gilruth at Langley wrangled a transfer to Muroc Dry Lake in the Mojave Desert of California, where NACA was beginning to experiment with high-performance air-planes. His name was Walt Williams, and he went to the desert to become an expert in the operations end of flying men in things that went real fast.

Another young man, only twenty when he came out of Virginia Polytech-nic Institute with an aeronautical engineering degree in 1944, caught the attention of people at NACA and was invited to Langley. Christopher Co-lumbus Kraft, Jr., was an impressive lad and they assigned him to the Flight Research Division. With the war in its final horrific year, he helped solve stability problems with the P-47 Thunderbolt, then the P-51 Mustang. Kraft's work on the last of the World War II fighters was enough to get him assigned as project engineer on America's first jet fighter, the P-80 Shooting Star. His boyish face made him an unlikely candidate for such an important post. But he had a commanding presence, and his understanding of aero-nautics was intuitive. From the beginning, Chris Kraft knew how to soak up information and make decisions. Sometimes he had to react in the few seconds remaining between life and death for a test pilot in trouble. He didn't work directly for Bob Gilruth, but the senior man—all of thirty-one years old himself—took note and was impressed.

Another fellow, a short little guy from Louisiana State University named

Max Faget (pronounced *fah-zhay*), showed up at the end of the war with a Cajun friend, Guy Thibodeaux. The two veterans took an LSU professor's suggestion and headed cross-country in a worn-out car to interview for jobs at the National Advisory Committee on Aeronautics at Langley, Virginia. While the nation sank into a deep postwar recession and returning soldiers found little demand for their skills as killer riflemen or tank drivers or mechanics, Faget and Thibodeaux blundered into lifetime careers. Both were offered jobs helping the flight researchers move into the far horizons of aviation.

The lessons of World War II rocketry were not lost on NACA or on Bob Gilruth. He'd seen the future and put the old-fashioned kind of airplane testing behind him. Now as chief of the new Pilotless Aircraft Research Division (PARD), he plunged—or perhaps soared—into the new world of rocket-powered flight. Two of his early acolytes, assigned to PARD as their first jobs, were Max Faget and Guy Thibodeaux.

Pilotless aircraft. The phrase was a euphemism. NACA's charter was in aeronautics. Its job was aircraft research, everything from designing and perfecting new shapes for wings to sorting out the structural and flight handling problems of airplanes that seemed always to go faster, go higher, and become more difficult for a pilot to control. When Max Faget looked around, his first question was how they were supposed to study aircraft at the next level of speed and performance. "There were no transonic airplanes in 1946," he said, "so how do we study something that doesn't exist? The old answer was to use wind tunnels, but there were no wind tunnels that could move air that fast either."

Gilruth's answer was to take NACA into the emerging field of rocketry. He put his new team to work designing one-tenth-scale airplanes that would be powered with air-to-air rockets left over from the war. "We had both American and British air-to-air rockets, a good stockpile of them stored out in bunkers," Faget said. "So we built model airplanes and flew them under rocket power."

When Bell Aircraft designed the rocket-powered X-1 aircraft, it used data from Gilruth's PARD flights. In their first designs, Bell engineers worried about strength and structural integrity. The X-1 would challenge the sound barrier, and they worried that transonic turbulence could rip off a wing. Gilruth looked at the thick wing they proposed and said, "No, you're wrong. You need a thin wing." He showed them the data from his P-51 tests, which he'd devised to help Chris Kraft. Gilruth's wing went against all conventional thinking. But the Bell engineers couldn't argue with his data. They changed their design. The X-1 rocket ship they loaded onto a truck in mid-1947 and hauled across the continent from Buffalo, New York, to California had a thin wing.

It didn't take long before the X-1 was being dropped from a B-29 over Muroc Dry Lake. Its pilot was a hotshot war hero from West Virginia who exceeded the test plan on the first flight by taking the X-1 to Mach .85, 85 percent of the speed of sound. Walt Williams was waiting with a warning when Chuck Yeager landed. The NACA wind tunnels only reached Mach .85, and if Yeager went faster, he'd be on his own. Even Bob Gilruth's P-51 version of an infinite wind tunnel couldn't do much better. Yeager wasn't worried. In the next seven flights, he and the test conductors ramped the X-1 speed up to more than Mach .90. On October 14, 1947, only six weeks after that first flight, Yeager went up again. The goal this time was Mach .97.

But it just felt too good. Yeager's chase pilot was Bob Hoover in a P-80 that had been smoothed and refined by Chris Kraft and his engineering team back at Langley. The P-80 was no match for the X-1. Yeager punched it into the transonic regime at Mach .97, marveled that it felt so slick, and blew through the sound barrier until he ran out of fuel at Mach 1.07. A new age of flight was born. Man had flown faster than sound on the back of a rocket, and neither aeronautics nor rocketry could rebottle the genie.

The fame went to Chuck Yeager. Nobody noticed the men behind the scene—Bob Gilruth, Walt Williams, Chris Kraft, and a few more. They didn't care. They were still learning. They just didn't know where it would lead.

The propeller was on its way out, and jets ruled in the military forces. Chris Kraft grew in reputation and expertise as NACA's point man on one jet aircraft program after another. Bob Gilruth's people turned their attention more and more to rockets. The little air-to-air rockets began to seem like toys. But now in the late 1940s and into the 1950s, rocketry was a rapidly advancing field of its own. Gilruth moved on to bigger model airplanes powered by bigger rockets, then began to focus on the rockets themselves.

"People were worried about incoming missiles and warheads," Max Faget said. "The big issue was aerodynamic heating, so we started to design and test our own missiles to check out heat loads."

They just did it backward. Wernher von Braun was firing modified V-2 rockets at White Sands, New Mexico, not far from El Paso. Other rockets were being designed, but none of the bigger ones was available to PARD. Bob Gilruth thought about it and one day gathered his team. They wanted data on something going through the air very fast, he said. Does it matter whether that thing is going up or down? "It was so obvious," Faget said. "Gilruth had access to some high-altitude bombers, so we started making test models, thin airfoils, but very heavy. We'd drop them like bombs on Wallops Island and they'd get Mach 1.1 or 1.2 on the way down."

They outfitted the models with sensors and little radios, and the telemetry data they recorded let them measure velocity versus time. That gave them a way to calculate drag and other details. "So Gilruth was getting data when nobody else knew how to get it," Faget said. "And later on, it was an easy step from going downhill with airfoil bombs to going uphill with rockets."

They flew more than a thousand drops over Wallops Island, Virginia, building a huge database on the cheap. Most flights cost only a few thousand dollars. Now and then, an elaborate test would cost $40,000 to $50,000. Among those were airfoil bombs that reached Mach 1.5, something no one before Gilruth and his men had ever done. Their data were unique. For its money, the nation bought an education in high-speed flight and reentry mechanics from a group of men who designed their curriculum on the fly. "If you'd decided to train some engineers in the basics of spaceflight," Faget said, "you couldn't have a better school than the one we went through. Every engineer in PARD learned to be a program manager and more than half of them went on to run some kind of big program."

Five decades later Max Faget would look back and marvel. "For us, the space race was Camelot," he said. "And in PARD, we were getting trained for Camelot."

Bob Gilruth, now balding and approaching forty, was rich cream rising toward the top of the National Advisory Committee for Aeronautics. In 1952, he became assistant director of research at NACA and began to consolidate his power and authority. "They must have thought I was the best guy for the job," Gilruth said with unaccustomed vanity, "and I happened to agree with them."

His expertise also was being sought at higher levels. He sat on the U.S. Air Force's Scientific Advisory Board and on the Planning Consultants for the Guided Missile Committee of the Joint Chiefs of Staff. In 1952, those were powerful advisory groups that were helping the American military move into the new era of high-tech warfare. When President Eisenhower abolished the committees during his Pentagon reorganization a year later, Gilruth found himself appointed to other committees. His vision of science and technology, and of airplanes and rockets, was clearer than that of most men of the fifties, and his guidance was sought. He was an astute manager too, and nearly everyone who worked for him at Langley was intensely loyal. He returned that loyalty in kind. Thus Bob Gilruth became one of those people whose names and faces were known to people in power, at least in certain narrow corridors of power, but who remained completely unknown to the world in general. But on the job, he got things done.

Gilruth managed by the "Oh, god!" technique. He'd hit his mostly bald head when he got bad news, *slap, slam,* and he'd say "Oh, god!" Everybody in the room would be seized by the thought that "we've got to help Bob out," and they'd go into long discussions of how to do this, how to do that, how to fix the problem, no matter what it was. He surrounded himself with the best people, and the fact that his head-slapping inspired additional loyalty didn't hurt a bit.

He was neither a glib speaker nor a particularly gifted writer. His technical reports, concise and filled with brilliant conclusions and the data to back them up, were the limit of his literary skills. Later in life he thought about writing his memoirs, but gave up after just fifteen handwritten pages. He was barely six feet tall and passed unnoticed in most crowds. His plain face, blue eyes, and slightly slouched posture gave him the look of a kindly grandfather, even in early middle age, rather than the look of a heroic American figure. He was uncomfortable at a dais and spoke mechanically. The emotion he felt seldom came through, and his audiences would listen respectfully, but soon there would be shifting bodies and glances at wristwatches.

Bob Gilruth was no Wernher von Braun, not in the faces the two pioneers showed to the public. Von Braun captivated audiences with soaring word pictures and legendary tales, holding them spellbound with either a prepared text or extemporaneous comments. He wrote as gracefully in English as he spoke, better than many of the time who called themselves writers. And he was an imposing figure with thick graying hair, a sturdy physique, and body language that radiated confidence from every pore of his 5'11" frame.

Whether Sergei Korolev might have been like either von Braun or Gilruth will never be known. He gave no public speeches. He wrote neither books nor articles for national magazines. He was received in the Kremlin as an honored guest, but there were no press conferences, and if photographers were present, their film disappeared into some Soviet limbo. Yet these three men in the 1950s, as different as different can be, were converging on a common point.

Wernher von Braun was a star. Robert R. Gilruth was a shadow. Sergei Korolev was a secret person.

And they needed each other desperately.

Von Braun's rocket unleashed a flood of launches, successful and not. Three days after Explorer 1, the Russians tried again with a Sputnik. It was their first failure. The rocket lost power, and when it fell back to Earth, Soviet Air Force pilots were sent out to find the satellite. It was recovered and some of its instruments salvaged for another try. None of this was reported. The abortive mission was hidden in Soviet archives for decades.

Then Vanguard continued its string of flopniks when its control system failed halfway to orbit February 5, 1958. And von Braun's reputation lost some glitter when his next Jupiter-C almost got to orbit on March 5, then fell back when its fourth stage refused to ignite. The Explorer 2 satellite it carried was lost. The American failures came in the bright light of press coverage and further shredded U.S. hopes and prestige with the shrapnel of imperfect technology. Finally, Vanguard put a tiny three-pound satellite into orbit on March 17. Its rudimentary photo cells transmitted fuzzy pictures of cloud cover and showed that Earth is slightly pear-shaped. It was the feeble grandfather of weather satellites, but for its time it was a marvel and Americans cheered.

Von Braun redeemed himself nine days later by adding Explorer 3 to the constellation of artificial orbiters. But near the end of April, Americans blushed again when another Vanguard blew fiery holes in the atmosphere and went into the logs as "failed to orbit." For every success, it seemed, there were two failures. And so far as the world knew, the Russians still had a perfect record.

A new Sputnik 3, another Soviet behemoth, hammered home that point when it reached orbit May 15 and began transmitting a wealth of data from its instruments. Less than two weeks later a modified Vanguard went awry during third-stage firing and put its satellite into the Atlantic Ocean.

The Chief Designer knew that he had America on the propaganda ropes and made a bold move for the *coup de grace*. By Soviet standards, the payload atop his next rocket was small, less than eight hundred pounds. But its destination was monumental: the moon. It launched June 25, 1958, from Tyuratam, almost immediately lost power, and became another of those failures that wouldn't be revealed for nearly three decades. The next day another Vanguard found a saltwater grave when its second stage abruptly quit downrange from Cape Canaveral. That failure was known to the world immediately.

By midyear 1958, the space race was in full thrust. The Russians and the Americans had tried fourteen times to put satellites into space. Six of them made it. Sputniks 1 and 2 lasted three months and five months respectively before tracing burning reentry paths back to Earth. Explorer 3's steep return in fire came June 28. Two American and one Russian satellite remained in orbit. The biggest by far was Sputnik 3, and Nikita Khrushchev boasted that "America sleeps under a Soviet moon." He didn't mention Sergei Korolev's failures but made much of the six times American rockets had blown up on the launch pad or tracked trajectories into the Atlantic Ocean instead of outer space.

In the terrestrial race for the hearts, minds, and admiration of Third World countries, the Soviet Union claimed a commanding lead. Where developing nations had looked to the West for guidance and role models, now they began to send their students to Moscow for higher education and the indoctrination that went with it.

Across the forty-eight states of America, the media riot battered at the gates of cherished institutions. In a half-breath of history, Americans began to doubt themselves. A country founded by brash, arrogant, and utterly self-confident people reeled with the fear that it might not be the best. Education came under harsh criticism for not producing enough scientists and engineers. The military's ability to protect the country from nuclear war was questioned. A missile gap created from whole cloth, and pounced on for political gain by Lyndon Johnson and Hubert Humphrey and later a young senator named Jack Kennedy, gave Americans to believe that the blustery Khrushchev could attack as soon as tomorrow morning.

The press and the Democrats raked the Eisenhower administration unmercifully for being too timid in pursuing spaceflight, for giving away the nation's superiority in missiles and nuclear arms, for underestimating the brilliance and technology of the Russians. The keening and whining mixed truth with half-truth and untruth.

The argument in the White House ran hot over who should run America's space program. The Army put the first satellite up. The Navy was second. The Army came back with number three. The Air Force was concentrating on war rockets but wanted a piece of the space action. Eisenhower listened to them all. A joint space command could settle the issue. But that was not acceptable to the scientific community. Give the military control of space, they argued, and the demands of secrecy would overwhelm the need to share data and to publish results and discoveries.

That was the dilemma. President Eisenhower still clung to his belief that if the United States was in space, it should be in the forefront of promoting space sciences. He insisted that the country must not appear to be overly militaristic. He also understood that satellites equipped with cameras could do things that no airplane could ever do. So while he spoke publicly of science in space, he spoke passionately in the secret meetings about surveillance from space. Eisenhower wanted spy cameras orbiting over the Soviet Union, to replace the U-2 airplane flights that he covertly approved. Sooner or later, he feared, a U-2 could get caught. But a satellite in orbit could not be touched. The question of how to have his cake and eat it too was solved by Richard Nixon.

Create a new civilian space agency, the vice president proposed. Give it responsibility for the science missions and for any other peaceful application of spaceflight. And let the military do its own thing. The new agency could be built on an old one, widely respected and already staffed with people who knew a few things about flight. NACA, Nixon suggested, could become NASA—the National Aeronautics and Space Administration. Some people and programs from the Naval Research Laboratory, the Jet Propulsion Laboratory, and the Army Ballistic Missile Agency might be transferred over to civilian control. And the U.S. Air Force could become the focal point, though not exclusively, for military space research.

Eisenhower liked the idea. He sounded out key congressional leaders and they liked the idea too. So did the scientific community. And most important, so did the leaders of NACA, who saw that their influence and power were about to be expanded exponentially. Paul Dembling, NACA's chief counsel—there is no such thing as a federal agency that does not have a legal staff—assembled a small team, including Bureau of the Budget people, to begin drafting the National Aeronautics and Space Act of 1958. Back at Langley, NACA already had people nibbling at the fringes of space.

NACA's current chief, Hugh L. Dryden, looked around and picked a mostly bald fellow now well into middle age to take on the role of space boss. Bob Gilruth thought about it no more than a millisecond before accepting the assignment. Through the spring of 1958, even before Congress chewed on the draft legislation and argued the merits of civilian control of space against military control, Gilruth started to get things together. Max Faget was never far from his right hand.

After his first week's experience at NACA in 1937, when he sat at a desk with nothing to do, Gilruth never again took kindly to waiting for instructions. He kept his people busy, and when he himself lacked guidance from above, he filled the vacuum by doing what he thought needed to be done. He understood intuitively that it was better to ask forgiveness than to ask permission. Faget learned by watching Gilruth. He and Gilruth discussed the inevitability of sending men into space, and Gilruth gave him the nod to look more seriously at the idea. Then Gilruth took the precaution of mentioning the idea in Washington and came away with a nod of his own to see what could be done.

Max Faget began sitting down with another PARD visionary, a self-educated engineer named Caldwell Johnson, and with Charles Mathews, a bright engineer from the Flight Research Division. "It was fate," Faget said, "three of us in the right place at the right time, with Bob Gilruth telling us to 'go get 'em.' "

In March and April 1958, the core of a manned space program was already formed. They sketched out a blunt-ended capsule that tapered toward the other end and could carry one man. They figured out how to attach the capsule to the top of a rocket. Faget's experience in running Gilruth's hypersonic heating experiments told him that the blunt end should be capable of absorbing and deflecting the white heat of reentry from space. Johnson and Mathews designed ways to add a cluster of retrorockets behind the heat shield. The shield itself should detach high in the atmosphere so that parachutes deployed from the other end would have less weight to lower to a landing—a splashdown—in the ocean. They called their capsule a "can" and designed it to sit atop one of the Air Force's Atlas intercontinental ballistic missiles. Then they started to build one. A civilian space agency still didn't exist, but by August, Faget, Johnson, and Mathews were dropping scale models of their can from cranes into the water and parachuting others into Chesapeake Bay.

The civilian agency soon legislated by Congress was elevated to administration status, making it nearly equal to the military services. Eisenhower insisted that a man-in-space program be given to NASA, not the military. He did give in to establishing a National Aeronautics and Space Council that would oversee civilian and military efforts, but only after making certain that the president himself was its chairman. In late July, he signed the bill that would create the National Aeronautics and Space Administration on October 1, 1958. He named T. Keith Glennan, president of Case Institute of Technology in Cleveland, as NASA administrator. Bob Gilruth was clearly destined to be in the new space agency's top management. Gilruth knew it. Faget knew it. Everyone in their inner circle knew it.

The previous fall, Faget had become unhappy with the smallish military and scientific rockets available for their PARD testing. Now with the dream of putting men into space becoming a national priority, Faget looked around and was equally unhappy with the rockets available for the kinds of research waiting to be done. Von Braun's Jupiters and Redstones were tightly controlled by the Army. The Navy's Vanguard had that nasty habit of blowing up on the launch pad or forgetting where it was going and opting for the ocean instead of orbit. Faget put an idea to Gilruth and to no one's surprise got a green light.

"We designed our own rocket," Faget said. "I got a team together and we designed the Little Joe rocket from scratch. Bob Gilruth backed us." Little Joe was for research. It would launch a full-size man-can to one hundred miles of altitude, then let it slam back through the atmosphere at maximum reentry velocity. By the day NASA officially came into being, Max Faget was ordering parts for the Little Joe rocket and Bob Gilruth was looking skyward at night with a new thought in his mind. Maybe it was time to make it all official and to add some new men to the team.

What was that word *Time* used a few years ago?

Astronauts. We need some astronauts.

3

Five days after the NASA ship set sail on its infinite voyage, they made it official. Man was going into space. Bob Gilruth would put them there. It was October 5, 1958.

At forty-six, Gilruth was now in command of the new Space Task Group (STG) and a program called Mercury. He was caught up in the grandest adventure of the twentieth century, an unlikely knight now leading the quest that he'd been on, mostly unknowingly, since 1937. He quickly set to the task of forming his team and finding out how to climb straight and high to the next frontier. NASA administrator T. Keith Glennan told him that he could have anybody he wanted, so he took the best, starting with the people he knew. He handed them responsibility, and the power to make it happen.

He gave Max Faget the Flight Systems Division, where spacecraft were to be designed and tested. Faget's assistant chief was Robert O. Piland. Neither of them yet knew much about recovering men from orbit. But they learned. Chuck Mathews got the Operations Division. His job was to run the missions. His assistant chiefs were G. Merritt Preston, destined to be the expert in launch operations, and Christopher C. Kraft, Jr., who became the legendary "man called Flight" in mission control rooms. Caldwell Johnson didn't have the seniority to run a division but was put in charge of the engineering branch in the new Engineering and Contract Administration Division. He soon had a score of degreed engineers reporting to him as they ran herd on the aerospace companies hired to build Mercury's hardware.

There were others scattered here and there through the new Space Task Group who became important men in the space race—Glynn Lunney, John

Hodge, Bill Bland, Aleck Bond among them. And there were some yet to be hired. Gene Kranz was not there on the ground floor but quickly found his way to Langley. Guy Thibodeaux was still needed on other work but would be swept into the Space Task Group when his growing expertise in rocket propulsion was needed. Walt Williams remained out in the California desert, where Muroc Dry Lake was now Edwards Air Force Base and the world's hottest airplanes traced vaporous contrails over the white desert floor. Williams had twelve years under his belt as operations boss and was charged with testing the still-to-be-flown X-15 rocket plane; Gilruth had something in mind for him, but not just yet.

One character would be brought into the fold in April 1959 but not report for duty until October. He was Air Force Lieutenant Colonel John "Shorty" Powers. Before long, Shorty's throaty baritone voice was as familiar to the world as any radio or television star's. He even added a new phrase to the common language: "A-OK."

Then there was that huge void in Gilruth's new organization. His job was to put men into space. He'd gladly go himself and have a score of volunteers like Max Faget lined up for their turn. But that wasn't the way to run a space program. Bob Gilruth needed specialists. Test subjects. He needed a group of men who were . . . what? Scientists? Engineers? Physicians? Pilots? Whatever their profession, they better be smart, brave, and in good physical condition. And given Eisenhower's insistence that America's man-in-space program be nonmilitary, maybe they should be civilians.

He'd put some people to work on that immediately, he decided. Because one thing was certain: Bob Gilruth needed those astronauts.

The nit-picking question was what to call them. Bob Gilruth favored *astronaut*. It had a nice ring to it.

Dr. Hugh Dryden, deputy administrator of NASA, wanted *cosmonaut*. "Astro" referred to stars, and Dryden argued that they were just at the beginning, exploring the cosmos, not the stars. So they passed the question around and Dryden lost. "Everyone we talked to seemed to prefer *astronaut*," Gilruth wrote in an unpublished memoir, "and this was the name that stuck. That is fortunate, I believe, because now when we say *astronaut*, we know we mean Americans, and when we say *cosmonaut*, we know we mean Russians."

That question settled, they got a surprise from President Eisenhower. They put together an astronaut plan, and NASA administrator Glennan sent it to the White House science adviser, James R. Killian. The idea was to select six men for the positions of Research Astronaut–Candidate. They'd get civil service positions in grades GS-12 to 15, making their pay between $8,330 and $12,770 a year, a decent salary for the time. The six-page draft, dated December 22, 1958, made it clear that "the Manned Satellite Project" was not looking for the kind of men who tested airplanes to the limits of endurance. Mostly NASA wanted observers who would do what they were told.

"[The astronaut] will contribute to the reliability of the system by monitoring the cabin environment, and by making necessary adjustments," the draft said. *So he had to be able to read gauges and twiddle the knobs.*

"He will have continuous displays of his position and attitude and other instrument readings, and will have the capability of operating the reaction controls, and of initiating the capsule from orbit." *More gauges to read, plus smart enough to point the capsule right or left or up or down, and even more important, smart enough to push the button that would fire the retrorockets.*

"He will contribute to the operation of the communication system." *He will talk into a microphone.*

"In addition, the astronaut will make research observations that cannot be made by instruments; these include physiological, astronomical, and meteorological observations." *At last. Something that may require a college degree, or at least a brain.*

". . . applicants must have successfully completed a standard 4-year or longer professional curriculum in an accredited college or university . . ." *Yes, a college degree.*

The candidate status of the selected men would last up to a year. After that, "As these men become proficient in the field, they will become eligible for Research Astronaut positions with salaries commensurate with those of the most highly skilled NASA Research Pilots and Aeronautical and Space Scientists." A supporting memo written a few weeks earlier envisioned perhaps three hundred applicants, ranging from physicians and scientists to airline and private pilots. The president of the United States took one look at the requirements for a Research Astronaut–Candidate and all but laughed.

"Tell them to pick their astronauts from military test pilots," he ordered. "We've got plenty to choose from and they're already on the payroll."

Some of the requirements for getting hired on as an astronaut were not changed. Maximum height was 5′ 11″. The Mercury spacecraft sketched out by Max Faget, Caldwell Johnson, and Chuck Mathews was seventy-four inches across its base. Anyone taller than 5′ 11″and wearing a bulky space suit wouldn't fit inside. The upper limit for weight was set at 180 pounds. Weight would forever be a concern in the space race, at least on the American side, because it took rocket power and heavy fuel to move a payload from the ground at Cape Canaveral to orbital space. The Russians seemed to have rocket power to spare. The Americans fretted over every ounce. That concern began with the weight of an astronaut and reached down to the weight of an electrical switch or wire. How grandly those worries would pay off; while the Russians were content to order off-the-shelf parts for their satellites and spacecraft, the Americans developed miniaturized and micro-miniaturized electronics that revolutionized or at least changed virtually every aspect of modern life. And it started with the 180-pound weight limit for astronauts.

There was an age limit too. Astronauts had to be at least twenty-five but no older than forty. When the search shifted to the corps of test pilots, it meant that NASA would be getting relatively senior and experienced men. It also meant that astronauts would be more than knob-twirlers and button-pushers. They'd be men who were accustomed to speaking their piece and taking control. What nobody realized was that they'd also be heroes.

The word went out to the military services on the first working day of January 1959. Commanders were asked to nominate test pilots who might become astronauts and to report back in a matter of weeks. The request was classified "secret." A Russian space launch that same day was not.

There was a subset of the space race being run that the public knew nothing about. It was a race to the moon. The Chief Designer had tried three times to hit the moon in 1958. The United States tried four. All seven failed. Korolev tried first, in June. His rocket did not reach orbital speed, fell back, and crashed. They said nothing about it.

Soviet policy was to announce success and conceal failure. The successes got names; the failures were anonymous. It was a shrewd policy. The world had the impression that every time the Soviets launched something toward space, they got it right. Sergei Korolev and the inner circle of the Soviet government knew better. There were more failures than successes. But each failure was a teacher. Spacemen of every nationality learned from their mistakes.

In August 1958, the first-stage failure of an American Thor-Able carrying a Pioneer moon probe did make news. The moon was still out of reach. But only the American failure made headlines. Korolev tried again in September and quietly failed. Again his rocket lost thrust and crashed.

Finally Pioneer 1 seemed to change America's luck. Launched in October, it aimed for the moon, lost power, and fell back after setting a new altitude record of 70,746 miles. *Nice try.* Sergei Korolev breathed a sigh of relief and got ready to go again. Before he could prepare another rocket and payload, though, Pioneer 2 lifted off the pad before reporters' eyes in November. The Thor-Able rocket and moon probe disappeared into the sky. It was just as well. After a long silence, loudspeakers blared the news that the third stage didn't work. The moon probe was probing the mud at the bottom of the Atlantic Ocean. America's record in the moon race it didn't know it was running was 0 for 3. And everybody knew those numbers.

The Chief Designer was delighted. His next moon probe lifted off in December. A half hour later he was not delighted. It was another failure. Again the world heard nothing about it. Two days later the United States launched Pioneer 3. It was a near-repeat of Pioneer 1. But this time its instruments discovered an outer radiation belt, beyond the Van Allen Belts, before the probe fell back to Earth. It was a partial success to the scientists involved and a complete failure to the rest of the world.

Lunik 1, launched January 2, 1959, was almost a success, not quite a failure. Korolev and his team all along wanted some way to show the world that their spacecraft was leaving Earth, not simply orbiting it. After discarding the idea of packing a small atom bomb in its nose cone that would explode on the lunar surface, they settled for a pyrotechnic container that released a yellow cloud of sodium in space at the point where the ship escaped from Earth. That put them in a mild bind. Astronomers across half the world would see the cloud. If they weren't forewarned, the cloud might cause international panic. So at the moment it was obvious that Lunik 1 had reached escape velocity and was on its way, the Soviet news service Tass put out the word.

The sodium cloud appeared on schedule over the Indian Ocean. Astronomers from the British Isles to Japan saw it and verified that once again, the Russians had achieved a space "first." Their spacecraft was on its way toward the moon. The Tass announcement prudently did not say that it would hit the moon, just that it was launched toward the moon.

Hitting the moon was the plan. Instead the trajectory was off and Lunik 1 missed by 3,728 miles. It sailed on by and became the first man-made object to go into orbit around the sun. So on one level, Lunik 1 was a failure. But that wasn't the way the Soviets played it. Lunik 1 was hailed for what it did, and if it didn't achieve its primary goal, it was an amazing achievement. The scientific community was delighted. Lunik 1 carried a magnetometer that showed for the first time that the moon had no magnetic field. Astronomers and astrophysicists had plenty to chew on as old theories about the moon's origin were resurrected and debated, new theories put forth. More than thirty years later, Soviet scientists admitted that Lunik 1 wasn't quite the success they had portrayed. By then, nobody cared.

Russian moons orbited the Earth. Now a Russian planet orbited the sun.

By late January 1959, NASA had a list of 324 men. Its own cadre of civilian test pilots, some of them like Joe Walker, Scott Crossfield, and a twenty-eight-year-old hotshot named Neil Armstrong in training out in California to fly the new X-15 rocket plane, were not among them. President Eisenhower told Gilruth to consider only military test pilots, and his people followed that order to the letter. They went through the files and quickly weeded out 216 of them.

The remaining 108 were divided into three arbitrary groups and invited to Washington. The teletype an Air Force captain received at Edwards Air Force Base was typical. It ordered him to report to a certain place in Washington, D.C., in late January 1959. The teletype was headed, in capital letters, TOP SECRET. The captain was ordered to wear civilian clothes and to discuss the teletype with no one. Since he didn't know what it was all about, there wasn't much to discuss other than to tell his wife that he was going to Washington and beyond that, he'd let her know.

On the appointed day, a roomful of test pilots got a secret briefing on why they'd been singled out. The new National Aeronautics and Space Administration was looking for volunteers to become astronauts. Before anyone could volunteer, they had to pass some tests. After the first two groups had been interviewed, sixty-four men were left. Five had been sent home for being too tall or too heavy (their records were not accurate), but still, NASA's Human Factors people decided they had a big enough group. The remaining thirty-nine test pilots were not invited to Washington and never knew that they'd even been on the list. Their names remained secret.

NASA asked the sixty-four to volunteer for further testing which could lead to being strapped into a can smaller than the driver's seat of a sports car. Once in the can, these test pilots were told, they'd be hurled into the airless vacuum of space by a rocket whose brethren frequently exploded within sight of land, and whose controls wouldn't let them do more than decide which way to point their nose. It didn't sound like a good career move to the men who said "No thanks."

So thirty-two were left. In early February, the picking, probing, and prodding began. For the next seven weeks, Gilruth's team had one goal: examine these men to the limits of science, medicine, psychology, and their own endurance. In groups of six or seven, they arrived at the Lovelace Clinic of Aeronautical Medicine in Albuquerque for the most intensive medical exams of their lives.

It was excruciating and embarrassing. Their blood was drawn, more than once. Nurses collected urine and feces samples, sent them to the lab for detailed analysis. They ran on treadmills while doctors watched heart rate and blood pressure readings. Eyes, ears, nose, throat, and the other bodily orifices were looked at and looked into. They were hooked up to electrocardiograms that recorded their heart activity and to electroencephalograms that recorded their brain activity.

These were healthy men. They flew experimental aircraft as their profession and engaged in every sort of physical activity for hobbies and fun. None of them could possibly have flunked a normal flight physical. One of them flunked one of these tests, where the standards had been raised to new heights. He was sent home with the promise that nothing of this would appear in his personnel jacket, that his name would never be known, and that he could continue to be a test pilot.

That left thirty-one.

The next stop was Dayton, Ohio, home of Wright-Patterson Air Force Base. Wright-Pat is a major U.S. Air Force research center. One of its units was the Aeromedical Laboratory. By the time the thirty-one left the lab behind in March 1959, they barely considered themselves astronaut candidates. They thought of themselves as victims. "Bend 'em until they break" was the rule of the day in Dayton. If they couldn't be broken physically, then find a psychological quirk that could be exploited. Two psychologists were assigned to each candidate and virtually lived with him for a week while he suffered and endured. Their questions never ended. When the candidate was too involved to talk, the shrinks stood around and watched.

If the medical tests in Albuquerque were intrusive, the tests in Dayton were uncomfortable, painful, and embarrassing. In one, their mouths were taped around bite pieces attached to flexible tubes so that lung capacities and oxygen use could be measured. They ran on treadmills until even the thought of another step was beyond comprehension. They were put on shakers and vibrated; eyeballs and teeth suffered for hours afterward. One day they sat in a dry sauna and baked for two hours at 135 degrees. Another day it was sensory deprivation, closed up alone in a black and soundless room for hours. On another, they donned pressure suits and sat in an altitude chamber while technicians evacuated the air until they were at the equivalent of 65,000 feet. One day doctors ordered them to take off their shoes, then plunged their feet into buckets of ice water.

At almost every moment, their reactions were measured. Heart rate, blood pressure, and temperature data flowed from sensors into recording machines. The medics logged how long it took to get the readings up, how long they took to drop back to normal. It was dehumanizing; they were lab animals. But they endured.

The thirteen separate psychological tests were exhaustive. One of the candidates later wrote about the difficulty of answering the simple question "Who am I?" Twenty times. "I am a man; I am a Marine; I am a flier; I am a husband . . ." By the last few, he wrote, it was difficult to figure out exactly who he was. Another had a bit too much of the wry in him. When a psychologist asked him to describe a blank white card, he studied it carefully, then deadpanned, "It's upside down."

Some of the worst days came in front of the camera. Nothing about these men was left to conjecture. It was important, the medics argued, that when these dire effects befell the astronauts there be a complete, *absolutely complete,* medical record so they could compare the original man with the failed specimen that came back to Earth. Surviving photos show them standing stoic and naked, every detail of their fit bodies intrusively recorded for medical history. In the only concession to modesty, the published reports showed them with white blotches covering their faces and crotches. But it isn't hard to put names to the men in the stark black-and-white photos. The camera looked at each of them from every angle, in close-up and full-body. They even were forced to squat over a camera set to shoot straight up at their genitals and anus.

Nobody asked for copies to take home to the wife and kids. "Nothing is sacred anymore," one of them complained.

Whether misguided or for a good cause, the medics' reasons were inviolate. They were already arguing about whether man could function in space, where gravity was essentially zero. Some said that their blood would pool in their legs, depriving their brains of oxygen. Or their eyeballs would lose their shape, rendering them blind. Or their hearts and lungs would not function properly, their muscles would fail to respond when fine motor skills were needed, reflexes would disappear, they would not be able to eat or drink. The test pilots scoffed when they heard about the fears. To a man, they'd been in extreme conditions, sometimes life-threatening, during their flying careers. Most of them had experienced a few seconds of zero gravity at the top of an outside loop, or during emergencies, and none remembered any problem. They admitted that a few seconds didn't really equate to hours or days in space, but they did know their own minds and bodies. Zero gravity, they said, was no barrier to getting the job done.

Finally the tests were over. One by one, the battered and exhausted candidates went home to wait. All they knew now was that one day soon the phone would ring and the news would be either very bad or very, very good.

After Albuquerque, the Space Task Group had thirty-one average supermen to evaluate and weed down to six. The reports from Dayton made it clear that eighteen of them were better-than-average supermen. The thirteen other folders went in a pile on a side table. Four men now held the future of America's man-in-space program in their hands. Charles J. Donlan, Warren North, and Stanley White sat around a table at Langley, Virginia, and debated the merits of each of the eighteen. Donlan was from management; he was Bob Gilruth's assistant. North was both a test pilot and an engineer; he represented that faction. White was an Air Force lieutenant colonel, a flight surgeon, and headed STG's Life Systems branch under Max Faget.

Gilruth was the fourth man. He didn't sit in on many of the intense discussions, but in the end he would approve or disapprove the decisions. One by one, they weeded them out, and after Gilruth gave the nod, folders joined the growing stack of rejects.

By Thursday, April 2, 1959, only seven candidates remained. They were the best of the best, and Donlan, North, and White couldn't decide whom to cut. They called Gilruth and explained. He looked at the folders, thumbed them through, and shrugged. "Then we'll take all seven," he said.

The next morning Donlan began making the last calls. One by one, he found them. "We'd like you to join us here at Langley," he said formally. "Are you still willing to volunteer?" Five of them answered in strong voices: "Yes, sir!" Within a week, their names would be known across the land and around the world. The five who responded were Gus Grissom, Deke Slayton, John Glenn, Gordo Cooper, and Wally Schirra. Donlan couldn't get through to the other two. Scott Carpenter was offshore on a Navy carrier. His wife, Rene, took the call at home. She was a go-getter and recognized Donlan's name. "Did he get in?" she asked.

"Yes," Donlan said, "but I have to have formal acceptance from him."

"He'll take it! He'll take it!" Rene Carpenter screamed. No way was this opportunity going to slip away because Scott had the bad luck to be at sea until later in the day.

"I'm sure he will," Donlan answered, "but I have to hear it from him. Have him call me." Rene made certain that messages were waiting at every place where Scott might check in when the ship docked. Before the close of business that day, he'd talked to Donlan and made it official. Scott Carpenter wanted in.

The last one was another Navy guy, a commander who was out flying when Donlan called. He left a message, but it got tucked away and the flier went home without seeing it. Through the weekend, the military grapevine hummed. He heard that others had been called. It was a miserable weekend. By Sunday evening, he was more than depressed. He was mad. *The best damn test pilot in the country and they passed me over? How could it happen?* He vowed to get to the bottom of it in the morning.

Meanwhile messages were going out to the other twenty-four: "Sorry, but we encourage you to try again if we ever need more astronauts." None of the rejects liked what they heard, but two were particularly incensed and vowed to try again. One was Navy Lieutenant James Lovell, who had just gotten his first ride in a Corvette from one of the other candidates. "What do these things cost?" he asked as they squealed around a corner and his head snapped back from full-throttle acceleration.

The answer was predictable: "If you have to ask, you can't afford it."

The other determined reject was Navy Lieutenant Charles "Pete" Conrad, the candidate who told the shrink that the blank white card was upside down. He wondered if his flippancy had hurt his chances. Then he decided that, hey, he was who he was. *Fuck 'em if they can't take a joke!*

Early Monday morning a white '58 Corvette squealed into the parking lot at base ops at Virginia Beach Naval Air Station and its driver charged inside. "Any messages for me?" he demanded. A young enlisted man looked up and handed him a slip of paper. "This came in Friday, sir." The officer took one look, grabbed a phone, and put through the call to Charles Donlan. Moments later he was grinning the grin of victory.

Al Shepard was in too.

Life magazine's impeccable sources told it that seven men, not six, would become Mercury astronauts. It sent seven photographers to the press conference scheduled for 2 p.m. Thursday, April 9, 1959. Each had his instructions: "Single out one of them and convince him to let you follow him home, anywhere in the country. Get pictures of his wife, his kids, his house . . . you know what to do."

The seven were gathered together for the first time in Walt Bonney's office at NASA's new headquarters in Washington. Each had followed instructions and showed up in civilian attire. For most of them, it was an uncomfortable feeling. They were career military men, and a uniform was what they wore to work. Now they were told to pack away those uniforms and wear civvies. A few of them didn't have the fashion sense of a twelve-year-old, showing up on day one in plaid sports coats and bow ties. But they learned quickly.

Walt Bonney was NASA's top public affairs officer. One of his young assistants, Paul Haney, briefed the nervous astronauts on what to expect when they faced the press. "They won't just ask about your flying," Haney warned them. "They'll ask about your religion, do you go to church regularly? They'll ask about your wife and kids. What's your political affiliation? Were you a Boy Scout as a kid?"

Until that moment, the seven had been thinking about the excitement of their new jobs. But each had moved about in the military, most had flown combat in World War II or Korea, and they were used to getting new assignments. This one was just a bit more extreme. But now something new began to sink in. What they were about to face wasn't just extreme. It was terrifying. They weren't just reporting in to a new commander. They were reporting in to the most demanding, cantankerous, and stubborn press corps in the world. At 1 p.m., nobody out there knew their names. At 2 p.m., they would become public property.

After his briefing, Haney left them alone to chat and collect their thoughts. This was something entirely new to them, and to make it worse, they hardly knew each other. As the clock moved toward the moment, they settled into their own private thoughts. The common thought was a rationalization: *Hey, how bad can it be?* At precisely 1:59 p.m., Walt Bonney walked to the rostrum at the Dolley Madison House on Lafayette Square and faced a room packed with reporters, photographers, movie cameramen, and the inevitable gaggle of gawkers and gofers.

"Ladies and gentlemen, may I have your attention, please. The rules for this briefing are very simple. In about sixty seconds we will give the announcement that you have all been waiting for: the names of the seven volunteers who will become the Mercury Astronaut Team. Following the distribution of the kit . . . those of you who have p.m. deadline problems had better dash for your phones. We will have a ten- or twelve-minute break during which the gentlemen will be available for picture taking. There will be no talk, however. Then we will reconvene, hoping that the p.m. boys have done their file and come back and start the presentation and the Q&A.

"It is now two o'clock. Please distribute the press kits."

There was a mad crush as reporters lunged for the clerks and secretaries who attempted to hand out the stapled kits. In the bedlam and pandemonium, few noticed that seven nervous and average-looking men had walked out onto the stage. The astronauts looked at the press stampede under way and wondered again what their lives had suddenly become. Walt Bonney waited a stroke until the seven were in place. Then he leaned into the microphone: "Gentlemen, these are the astronaut volunteers. Take your pictures as you will, gentlemen."

Dozens of photographers, some shooting from fixed positions on a photo stand, others free-roaming, elbowed and jostled for position. A few hollered questions or instructions to the astronauts, which were ignored according to their instructions.

The seven *Life* photographers surged in with the rest of them. Ralph Morse elbowed his way to the front and turned his camera from one astronaut to another. He saw the bemused look on the face of the one near the middle, the thinning-haired reddish-blond who seemed a bit older than the rest, and concentrated on him. Whatever surprise the man had experienced was now replaced with curiosity at the spectacle of grown men scrambling for position, some of them using the long lenses of their cameras as weapons to knock a competitor on the shoulder or temple, forcing him to give away the vital inch that put a picture in perfect frame.

Morse tried to make eye contact with the redheaded man, but it was no use. He understood what was going through the minds up on the stage. Morse was the second combat photographer hired by *Life* in World War II. He landed with the Marines at Guadalcanal, was sunk in the Battle of Coral

Sea, photographed Jimmy Doolittle taking off from the carrier on his way to bomb Tokyo. Morse knew what it was like to operate in a pressure situation. When *Life* decided that he'd been risked enough in the Pacific, they brought him home, then sent him to England. His first assignment in the new theater was to wade ashore with the third wave at Normandy. He rode with Patton into Paris and photographed Jodl signing the German surrender. Back after the war, his life was not quite as exciting. He'd virtually lived for the first week with a young black named Jackie Robinson when he'd come up to the Dodgers and broken baseball's color line, and he'd enjoyed himself for ten years after as *Life*'s resident photographer on Broadway. Now he sensed that his career was changing again. He kept his eye on his chosen astronaut when Walt Bonney called the session back to order and let NASA administrator T. Keith Glennan make the introductions.

From left to right, the nation's Mercury astronauts were Air Force Captain Donald Kent Slayton; Navy Lieutenant Commander Alan Bartlett Shepard; Navy Lieutenant Commander Walter Marty Schirra; Air Force Captain Virgil Ivan Grissom; Marine Lieutenant Colonel John Herschel Glenn, Jr.; Air Force Captain Leroy Gordon Cooper, Jr.; and Navy Lieutenant Malcom Scott Carpenter. The supposedly jaded journalists in the auditorium broke into spontaneous and thunderous applause. The astronauts were shocked again. You could see the expression on their faces: *What have we gotten ourselves into?*

Their wives and families wondered the same thing. None expected the onslaught. Hometowns and addresses were included in the press kits handed out twelve minutes earlier. Reporters had run for the phones to dictate afternoon leads while the photographers jostled. With the bare-bones story ripped from the typewriters of rewrite men, the on-scene fellows read off the vital statistics. Before they hung up to get back to the Q&A, rewrite men had handed the addresses to editors and calls were going out. Within half an hour, reporters and photographers were descending on unprepared wives and kids, swarming their homes and demanding answers to a thousand too-personal questions.

It wasn't quite as bad for parents, but they got their share of reporters too. And over the next few days, teachers and preachers, classmates and bridge partners, fellow officers and pilots all found themselves being interviewed by reporters looking for the color and character of these Mercury astronauts and for the one obscure detail that would explain just exactly who and what they were.

Back in Washington, before he allowed the astronauts to be questioned, Walt Bonney introduced the key medical people involved in the selection, along with Bob Gilruth and Charley Donlan. He asked Gilruth to say a few words, and that's exactly what he got.

"Ladies and gentlemen," Gilruth began in his flat, unexciting voice, "I'm delighted to be here on this occasion. I am sure you are all familiar with the objectives of Project Mercury. I don't intend to make any speech whatsoever about it because I know you are all here primarily to meet our astronauts.

"Thank you very much."

That was the reporters' primary introduction to the man who would be America's most important space figure through the coming years, and his to them. Gilruth sat back, slightly embarrassed, and the reporters did their best to ignore the kindly looking bald man.

Each of the others said a few words before Bonney let the astronauts talk. Dr. Randy Lovelace got a few chuckles with his comment about the medical tests. "I just hope they never give *me* a physical examination," he said, then continued: "It has been a rough, long period that they have been through . . . I am not worried about their stability, their powers of observation, or their powers to accomplish the task . . ." Finally, Bonney turned to the astronauts and asked Scott Carpenter what his wife and kids thought about all this. Carpenter hesitated a moment. Like the rest, he'd never thought about facing an auditorium full of reporters waiting to write down his every word. He didn't give them many.

"They are all as enthusiastic about the program as I am," Carpenter said, and then shut up. He didn't sound enthusiastic at all.

Bonney suddenly realized that he had to do something to draw these guys out. "Suppose we go down the line . . . Has your good lady . . . had anything to say about this?" It wasn't Bonney's finest question, but things warmed up soon enough.

After Gordo Cooper echoed Carpenter's concise answer, it was John Glenn's turn. He took a breath, relaxed, looked calmly at the crowded room —photographers were still pushing and shoving down front—and began to make his reputation.

"I don't think any of us could have really gone on with something like this if we didn't have pretty good backing at home, really," he said. "My wife's attitude toward this has been the same as it has been all along through all my flying career. If it is what I want to do, she is behind it, and the kids are too, a hundred percent." Pencils scratched rapidly on notebooks. There was something in Glenn's demeanor, and in the way he put some thought into his answer, that marked him.

Grissom, Schirra, Shepard, and Slayton followed, the first two seeking some depth in their answers, but Shepard and Slayton answering tersely. Slayton was obviously the hard case of the seven. "What I do is pretty much my business, profession-wise," he said. "My wife goes along with it."

They were asked about motivation, and Slayton surprised the reporters with a measure of eloquence. "I feel that this is the future, of not only the country but the world," he said with feeling. "We have gone about as far as we can on this globe, and we will have to start looking around a bit."

Again Glenn took the higher-still road. "I got on this project because it probably would be the nearest to Heaven I will ever get and I wanted to make the most of it," he said, and shared chuckles with reporters. "But my feelings are that this whole project with regard to space sort of stands with us now as . . . the Wright brothers stood at Kitty Hawk about fifty years ago, with Orville and Wilbur pitching a coin to see who was going to push the other one off the hill down there. I think we stand on the verge of something as big and as expansive as that was fifty years ago."

The question of who would be first into space was handled by Gilruth. He explained that designs were still incomplete and that much testing remained to be done on Mercury. "We are still a very long way from being far enough down the road to answer," he said. Sensing the dangers involved for the first time, a reporter asked if the astronauts were confident about coming back from outer space. All seven instantly raised their hands. When another reporter asked if they were ready to go into space *now,* all seven raised their hands again. Glenn raised both of his.

They were asked about religious faith, as Paul Haney had predicted. The first few answers were perfunctory. Carpenter went to church when he could, Cooper was a Methodist and didn't say how often he graced a pew. Then John Glenn stepped up and hit a moral home run. "I am a Presbyterian . . . and take my religion very seriously, as a matter of fact," he answered. "I have taught Sunday school . . . and I was on the board of trustees at the last duty station . . . We are very active in church work, and the kids are in Sunday school and all the things connected with the church work."

Now the reporters were scribbling furiously. They had a winner and they knew it. Ralph Morse aimed his camera and shot a full roll of 35-mm film as Glenn continued to speak. ". . . you will find a lot of pilots who like to take what I consider to be sort of a crutch and look at this thing completely from a fatalistic standpoint, that sometime I am going to die, so I can do anything I want in the meantime, and it doesn't make any difference because when my time comes, I am going anyway."

He paused, took a breath, then went on. "That is not what I believe . . . We are placed here with certain talents and capabilities. It is up to each of us to use those . . . as best you can . . . I think there is a power greater than any of us that will place the opportunities in our way, and if we use our talents properly, we will be living the kind of life we should live."

While he spoke, Shepard and Schirra and then slowly the others recognized a basic fact of their new life. Each of the seven was an intensely competitive man. Each got to this moment by winning out over others. And each desperately wanted to be America's first man in space. So as Glenn talked, a new competition formed before the eyes of an unaware press. For one of them to be first, six had to be beaten in the coming competition. They would become fast friends for life, these men who would be known as the Original Seven, and their allegiances to each other were already forming bonds of emotion and steel. But none of that would stop any of them from doing everything within the bounds of fairness to defeat these friends and make them sit and watch while *I ride that rocket toward the stars.*

And now John Glenn, through a couple of adroit answers to reporters' questions, had cast himself as the man to beat. *Hell, we haven't even gotten through our first day on the job and this marine jet jockey is already running away with the crowd.*

It was Wally Schirra's turn. He tried to cut into Glenn's lead. "I have followed the Episcopal Church as my faith," he said with a serious tone to his baritone voice. "I have been an active participant in church activities." That part out of the way, Schirra reached for the philosophic. "I think I should like to dwell more on the faith in what we have called the machine age. We have faith in the space age . . . All of us have had faith in mechanical objects. We have been driving cars, sometimes not successfully; we have been flying airplanes, some not successfully. I think the space program will have problems in it. Again, you have to believe that we are trying to *do something* and we are going to do it as well as we can."

Al Shepard glanced down the table at Schirra and Glenn, then dug up some eloquence of his own. He had to establish to this crowd that he had a mind of his own and some insight into the job that lay ahead. "I don't mean to slight the religious angle in my answer," he said with an earnest voice.

"The [Mercury] project is . . . merely one step in the evolution of space travel . . . It is also pertinent to state that the risk . . . is not as great as you possibly feel that it is." His voice grew stronger and was tinged with stridency. "I would like to discount the fact that this project is extremely hazardous," Shepard said. "It is not a technical race, it is a step in the evolution of space travel."

But it was a race and Shepard knew it. The humiliation each of them would feel if a Russian beat them into space . . . *No, that can't happen. It's unthinkable.*

They were asked again about their wives' reactions.

Grissom: "When I said something to my wife, what she thought about my volunteering, she said, 'Do you even have to ask?' "

Schirra: "I remember my own reaction, that we were a bunch of idiots. My wife felt the same way . . . [Now] I immediately say, 'If you will take . . . time to think about it . . . you will realize that this is something we are very serious about.' "

Shepard (forgetting that Glenn had yet to speak): "She was in complete support of my professional decision . . ."

Slayton (again the hard case): "My wife isn't too concerned about what I do professionally. She is more concerned about whether I can find a babysitter and whether there is a commissary nearby to buy groceries . . . Whether I am in this program or in a filling station doesn't make any difference."

Carpenter: ". . . my wife called Washington and volunteered for me."

Cooper: "My wife is also a pilot and is quite sympathetic . . . She is enthusiastic."

Then it was John Glenn's turn. Shepard winced inside at the way he'd blown his own answer when Glenn began to speak.

"My wife made a remark the other day, I 'have been out of this world for a long time, so I might as well go on out.'

"I think all the wives . . . when they first hear something like this, they have reservations about it because they don't know anything about it. It is like all of us . . . you are very interested, but you sure want to find out a lot more about it before you are willing to place your neck way out there.

"I think I can best describe my wife's reaction in that, as she has learned more about the program . . . she has become as enthusiastic as I am."

So John Glenn covered for Slayton and Shepard and Grissom and expanded on the others' thoughts. He wasn't only a man's man, he had empathy too.

And so it went. When Walt Bonney finally called an end to the press conference, there was another spontaneous round of applause from the reporters and sighs of relief from the seven astronauts onstage. It had been a stunning and revealing experience for them. At some point in the afternoon, each of them realized that somehow he'd gone from an anonymous guy to some kind of hero.

In a book he wrote later with Al Shepard, Deke Slayton marveled at the transformation: "It happened without us doing a damn thing. We show up for a news conference, and now we're the bravest men in the country. Talk about crazy!" It was about to get crazier. By the next morning, America knew their names, their faces, where they'd been born, and what they were going to do now.

Rene Carpenter woke up the next morning with reporters and television trucks parked out front. Later that day, more showed up and she brought the kids home from school for pictures. She told the reporters to make themselves at home. While she was gone, they moved the furniture out of her living room to convert it into a studio. She looked at the chaos and asked, "Do you mind if I take a minute to run a comb through my hair?"

"Sure, Mrs. Carpenter, down the hall and second door to your right."

Suddenly she realized what was happening. *A reporter showed me the bathroom in my own house!* Privacy was gone. It wasn't just the husbands who were public figures; it was all of them. "It's as if I'd been acting on a dark stage all my life," she laughed later, "and suddenly someone turns on the spotlight."

The publicity part was simple. What they had to do was to get out there and win the space race. That's what heroes do. And once again, America had heroes to adulate. Slayton was right. Before the seven could take a collective deep breath, they'd been mounted on the highest pedestals any of them had ever seen. To a man, they shook their heads and wondered what the hell it all meant.

Ralph Morse bulled and brashed his way to John Glenn's side and convinced him to let the *Life* photographer follow him home to suburban

Virginia. It was a career-making day for both of them. Morse spent a week in and around the Glenn home. His photos were the best of *Life*'s family shots. More important, he became a friend. Annie Glenn had a severe stuttering problem, which she later overcame through sheer determination and hard work, and Ralph handled it with tact and understanding. He made certain that *Life* didn't mention the affliction in its story, though eventually a crass reporter for the *New York Times* made too much of it in her story about Annie and hurt her badly.

When Morse returned to New York, he barged into the photo chief's office and made it clear that he wanted to be *Life*'s primary space program and astronaut photographer, no matter who else wanted the job or what it took to do the job right. He got his wish. As all of the astronauts and many NASA officials would learn in the decades ahead, when Ralph wanted something, it was next to impossible to say no. He hadn't walked through Guadalcanal and Normandy without a scratch for nothing.

Glenn had been assigned to a flight research desk at the Navy Aeronautics Bureau, in a bunch of old wood buildings about where the Vietnam Memorial would be built years later. Of the seven, he was the only one who didn't move his family to Langley. His kids were in good schools and Annie had her friends. They agreed that it would be better for John to stay in bachelor quarters at Langley and come home on weekends. "The space program wasn't guaranteed to continue," Glenn said. "We didn't know what was going to happen. We were making up the music as we went along."

Glenn was already a familiar face at Langley. Chris Kraft met him first in 1956 when Glenn was assigned to the Aeronautics Bureau's F8U desk and Kraft was working with the Navy on solving some in-flight problems with the jet. When NASA asked for volunteers to come down to Langley a day or two a week to try out newfangled hardware like hand controllers to replace the old control sticks, and to take part in simulations, Glenn quickly volunteered. Along the way, he and Kraft developed a respect for each other that would turn into genuine friendship.

Five thousand miles away, and several days later, the Chief Designer saw the press reports from London and the United States. One of his staff translated the newspaper stories and Sergei Korolev read them all. He didn't have any spacemen of his own yet, but he thought that it was time to push his own program in that direction. From the beginning, he'd looked toward the day when a man would crawl into a big Sputnik and be hurled into orbit. But his immediate attention had been focused farther out. In mid-1959, he would be back at the job of sending spacecraft toward the moon and rub the American noses in Soviet space supremacy one more time.

But meanwhile, those Americans were taking the right steps to put a man up there with the artificial moons. They couldn't be allowed to get there first. He needed cosmonauts. Such a move, even to begin asking for volunteers, needed approval from the very top, Nikita Khrushchev, maybe even the entire Soviet Presidium. It was time to start that particular ball rolling.

Astronaut training began to accelerate in the United States. They were on their own to keep fit, and they preferred the individual efforts. Some of them ran. Others worked out in the Langley gym. Strangely for competitive men, they had little interest in competitive activities like handball or racquetball. Some of them played a little, but it wasn't with the passion that many men experienced when a racquet or a toughened palm slams a ball into a corner.

They watched their diets. But what watching they did on their drinking, except for John Glenn, was watching the after-duty glass be filled and emptied and filled again. They were beginning to recognize what the reporters and the American public recognized from the start. They *were* a breed apart and it wasn't just because they flew experimental airplanes and accepted death as a working companion.

They could hold their liquor better than most men. Their reflexes were whiplike. They thought clearly and coolly under pressure. They didn't scare. Whether the day had a confrontation with danger or not, they liked to party, and if any reporter had been smart enough to recognize it from the way most of them referred to their wives, they also liked the ladies. The girl-in-every-port syndrome didn't just apply to sailors. Test pilots carried their own "books," and Al Shepard's was said to be the thickest of them all. The NASA men who traveled with him over the years claimed that they never hit a town where he didn't pull out his book with the comment "Well, let's see who we know in Tucson" (or St. Louis or London or anywhere else).

Some had known combat, shot down the enemy in Korea, and limped home with their own planes full of holes. All knew the excitement and moments of stark terror that went with being a test pilot. They'd been in harm's way and would go there again laughing at the devil and sweating to find a way to deprive him one more time. They flirted with death on the job and with the ladies in the evening. Each of them was ruggedly good-looking. They were hilarious, especially after a drink or two, and played practical jokes and told stories about fighting and flying that drew women close, and often kept them there for an evening or a night.

That was what happened to test pilots. What happened to astronauts was the stuff of legend, and most of it can't or won't be known outside of the inner ranks. But enough of it got out before the space race ended to add an aura of intense sexuality to their reputations. They didn't just have the right stuff. They had the magic like nobody had had it before.

John Glenn was the exception. When the partying got too much, it was Glenn who lectured the other six and talked about the image they owed to their country and to its people. Sometimes the others resented Glenn's Goodie Two-Shoes attitude. More often when he took them to the woodshed for carousing and womanizing, they listened and were contrite.

For a while.

Then the temptations, more in a week than most men experience in a lifetime, and of better quality too, would be too much and they'd be at it again. It was in their warrior blood, and maybe in their genes, and when they surrendered, it was unconditional.

Their wives knew it, of course. They'd long ago accepted what their husbands were. Maybe it was part of the attraction, not in a kinky way, but loving a superman and knowing that when he came home, it was to them. Certainly it wasn't resignation. To a woman, the wives of all astronauts were strong, intelligent, and mostly beautiful. A few succumbed in later years and became alcoholics. Most did not. Though the divorce rate of astronauts eventually far surpassed national averages, those events were a long time coming.

And through it all, never once did a hint of scandal touch John Glenn. He was all grown-up, but still a Boy Scout.

Who were these guys, anyway, who burst into the American consciousness on April 9, 1959, and with others who joined them in years ahead, held a nation in thrall for an entire decade? Each of them was a firstborn son. Firstborn sons often are superachievers, daredevils, men who take risks for a cause and frequently succeed. Unless they die or run afoul of the law. They all grew up in small towns, not an Angeleno, Chicagoan, or Bostonian among them. Each of them was married. Each had children. So on that level at least, they were typical American family men.

Five were college graduates. Scott Carpenter had failed twice to get a degree at the University of Colorado but told the Navy otherwise. Nobody checked until after he became an astronaut. Carpenter picked up more than enough education in the Navy to become a qualified engineer. He was a brave man, but with a flaw that all the shrinks missed. He had a poetic side, or maybe it was a sense of curious wonderment woven into his thirst for adventure. It didn't show up in all the tests, but when it did, it would cost him. John Glenn had dropped out of college to join the Marines and become a fighter pilot. The missing college degree was Glenn's only drawback. Because of their other qualities, NASA waived the college requirement for both Carpenter and Glenn.

The first astronauts were average men in size. Gus Grissom, at 5′ 7″, was the shortest. Al Shepard was 5′ 11″. John Glenn, Deke Slayton, and Carpenter were a half inch shorter. Wally Schirra stood 5′ 10″ and Gordo Cooper was 5′ 9″. Schirra and Glenn weighed 180 but would go into space weighing less. Cooper was a skinny 150, Grissom was a 155-pound fireplug, and the rest scaled in at 160.

None of them were dummies, despite a statement uttered in anger about one of them years later by a certain NASA official. But there were differences in intellect. Shepard was probably the smartest. Carpenter seemed to be dumber, or maybe it just looked that way because of his missing ration of common sense. The rest were bright and alert, and all of them ranked well above average on IQ and personality tests.

Al Shepard, thirty-five in 1959, was the one who always saw the big picture, and how the details would come together to make it work. He was extraordinarily observant, a major contributor whether in long, tedious meetings or in the adrenaline rush of postflight debriefings. He loved his white Corvette and would race anybody. Shepard's influence led to Corvette's becoming known as the astronaut car, and because they got some special deals, there were years when most astronauts, and some NASA officials, drove identical Vettes, a fact that would cause trouble one day.

At thirty-seven, John Glenn was the old man and understood human nature better than any; he instinctively knew that image was as important as action, but when he accepted the hero's mantle and played the role of internal censor, these were natural acts for him, not contrived. He flew in both World War II and Korea, five times earned the Distinguished Flying

Cross, and had nineteen Air Medals. In 1957, he set a transcontinental speed record, averaging supersonic speeds from coast to coast including in-flight refueling.

Gus Grissom, thirty-three, was shy and gruff. He hated the publicity and attention and would go to lengths to avoid reporters. He was uncomfortable in interviews and even more uncomfortable when he was forced to stand up in a public place and deliver a speech. But he was a great pilot, a consummate engineer, was quick to see when changes in a spacecraft's design would make it easier to operate in space. He won the DFC and Air Medal flying combat in Korea. He also was a ladies' man, so fervent in his pursuit of conquest that he once left a party to take his wife, Betty, to the airport, then returned with another young lovely on his arm.

Rumors surrounded Grissom, including the unproven rumor that he fathered an out-of-wedlock child born to a secretary at the McDonnell Aircraft Corporation in St. Louis. One story was not a rumor. There's a certain kind of small black fly that hatches in the spring around the space center south of Houston. Swarms of the bugs can splatter windshields, but their real distinction is that male and female catch each other in midair and fly along happily mated. Grissom told a *Life* magazine reporter that he envied those insects. "They do the two things I like best in life," he said, "flying and fucking—and they do them at the same time." For years thereafter, the insects were known as Grissom Bugs to local residents.

Deke Slayton, thirty-five, was a technician. He was the best engineer of the group and pushed his ideas on what was right and what was wrong with hardware. He argued for changes when he thought he was right, and since he usually was, the changes were made. He also had a temper. People who felt Slayton's wrath, from reporters to NASA and contractor engineers, knew the sting of his voice. But he also could be reasonable. On the rare occasions when he was wrong, he knew how to reverse course and do the right thing. He would become the most powerful of all astronauts.

Wally Schirra, thirty-six, was the joker. They all enjoyed a good laugh, but Schirra's penchant for practical jokes—he called them "gotchas"—bordered on genius. He was another superb pilot and he might have come by that genetically. His father was a barnstormer and his mother a wing-walker in the wild days of aviation. He was the third astronaut to wear the DFC and Air Medal for Korean combat. He had the touch in controlling a spacecraft; shifting its position by an inch or two while traveling at 17,500 miles per hour was a piece of cake to him. Schirra's competitive streak was a mile wide. He wanted to be first and best at everything. Because of that, he understood the space race and its implications. Like Al Shepard, he loved sports cars. One of the first he bought after becoming an astronaut was a bright yellow Austin-Healey. Later he bought Brigitte Bardot's Maserati and it wasn't unusual for drivers to look in their rearview mirror and see him coming up fast, sometimes more than one hundred miles per hour.

Scott Carpenter, thirty-three, was brave, almost too brave. His exploits in planes, cars, and spacecraft often exceeded the limits, and he was banged, bruised, or had broken bones more than he should. He didn't have the combat experience of Glenn, Grissom, or Schirra, nor as many test pilot hours as Slayton, Cooper, or Shepard. When his lack of a college degree was revealed, the University of Colorado gave him one after his Mercury flight. He did have physical stamina and moxie, and no matter his flying history, he was a terrific pilot.

Gordo Cooper was the kid, only thirty-two. His Oklahoma drawl some-time fooled people into thinking he was slow. That was a mistake. He'd let someone walk into the trap of underestimating his intellect or condescending to him at a party or in an intense engineering meeting. Then he'd spring the trap with an incisive comment, shrewd observation, or on-the-money recommendation. He took flying home with him and owned his own plane. His wife, Trudy, also was a pilot, with a flight instructor's rating, and their daughters grew up in the air.

Cooper also raced in sanctioned sports car events and liked to challenge Al Shepard in their identical Corvettes. Except they weren't identical. He didn't tell Shepard that he'd changed the gear ratios in his differential; when he left Shepard in the dust, the older astronaut assumed that it was Cooper's superior driving skills that put him into the lead. It wasn't. It was one of those gotchas they all played on each other. The difference here was that Cooper laughed in private and Shepard never knew the truth.

The womanizing was a problem, and no matter how often John Glenn lectured, the temptations simply overcame temporary good intentions. Through the early years, more than one woman bragged about bedding six of the Original Seven. Nobody ever claimed to have seduced John Glenn. The exploits of the others were an open secret in the bars and motels of Cocoa Beach. Reporters knew the stories but didn't write them. *"The public's right to know"* was a phrase that hadn't been invented in the late fifties and into the sixties. There was an unwritten gentlemen's agreement between reporters and the astronauts: If it didn't get entered on a police blotter, it wasn't a story.

"It was so wide open in the early days, I can recall seeing lines—three or four ladies in a line—outside the oceanside rooms of the good old boys." The words came from Paul Haney, whose public affairs assignment put him into frequent contact with astronauts. "I liked to stay back there too, just in case I got lucky."

In later years, the extracurricular romances got more serious. But some like Al Shepard were a long time in slowing down. Shepard was one who didn't need to trade on his astronaut status. Haney recalled being in an Arizona town where Shepard gave a speech. Later Haney picked up a local lass and went home with her. They'd no more than gotten comfortable on her couch when her roommate came home—with Shepard in tow.

"We pretended we didn't know each other," Haney remembered. *Oh, you're in the insurance business? I'm a salesman myself.*

One of the first to claim bragging rights on six of the seven was a Florida reporter nicknamed Wicky. Other reporters quickly dubbed the six as the Wicky Mouse Club. A few years later in Houston, a pretty brunette assigned to the public affairs office took her job literally and laid claim to the six, and a few more as new groups of astronauts came along. Haney had the embarrassing chore of driving back from a party with the sexy scorekeeper and one of the newer astronauts in the backseat.

"I had a four-door '62 Nash," Haney reported, "and I became aware that two male feet were digging upside down for traction on my inside roof. With all the appropriate groaning and grunting." If he'd known how famous that man's footprints would become, he might have saved the headliner from that old Nash.

It was all passed off under the heading of "boys will be boys." But the Original Seven were special. They came together as a team, as friends, and as competitors in April 1959. Then they changed the world and wrote their names in the stars.

4

While the seven got ready to transfer to NASA at Langley, Virginia, they pored over plans for the Mercury capsule and its systems, submitted to press interviews, and even spent a day in Washington at a closed-door congressional hearing. At the same time, another part of the Space Task Group found itself in a fight with the flight surgeons.

Chuck Mathews' Operations Division wrestled with the question of how to control, or at least keep track of, a capsule that circled the world every ninety minutes. The obvious solution—one denied to the Russians by their own policies of secrecy—was to build a series of tracking stations around the world.

Even then there would be gaps, long minutes when the capsule passed out of line-of-sight range at one station and before it appeared over the horizon at the next site. For longer flights, the gaps could be even longer. The reason was that a capsule in orbit is literally on its own. An airplane in the atmosphere is swept along as the Earth turns. A satellite above the atmosphere is not. The Earth both turns and wobbles below. So each time a Mercury capsule completed one turn around Earth, it would trace a new track on the ground. It would pass over the same spot only once every twenty-four hours. If it passed directly over a tracking station, then it would pass north or south of that station ninety minutes later. After a few turns around the world, it would pass too far north or south for that station to track it. The 70 percent of Earth covered by water complicated the problem. At least a few tracking ships were needed to cover broad expanses of ocean. They had the advantage of mobility; the ship could steam in the direction of the next pass and get a few extra minutes of coverage denied to ground stations.

None of that was good enough for the flight surgeons. They wanted full-time coverage, 100 percent of the capsule's orbit available to both radio and television reception. One of Mathews' assistants, a Canadian-American with a distinctly British accent named John Hodge, bore the brunt of the doctors' arguments. They were again playing the human-frailty card. If an astronaut got sick in orbit, or was injured, or was suffering from any of the space maladies they conjured up in their wildest fantasies, they wanted to offer treatment and advice instantly. John Hodge shook his head in dismay. The surgeons didn't understand. There simply wasn't enough money to do what they wanted.

A workable tracking system was budgeted at about $41 million. A full-time system would cost ten times that much—no, twenty or thirty times. In fact, because of the world's political realities, some countries would object to hosting any tracking site. Full-time was impossible. Dr. Stan White and a few others didn't care. They wanted it. Hodge thought it over from the medics' point of view. If an astronaut got sick, what could they do? The answer, he realized, fell somewhere between not much and nothing. He convened a meeting, brought in both engineers and doctors, and laid it on the line.

Hodge asked the obvious questions: A sick or injured astronaut had to be brought home, right? The flight surgeons agreed. It takes twenty minutes after retrofire for the capsule to drop through the atmosphere and land somewhere, right? The flight surgeons, caught in a situation where the laws of physics overruled the art of medicine, agreed again. So full-time tracking coverage doesn't matter, does it? Until the astronaut was back on Earth, there was nothing the flight surgeons could do anyway. Stan White and the rest recognized a corner when they were backed into it. They dropped their demand for a tracking network that was impossible anyway.

The man assigned to scout out sites was Hartley Soulé, the same Hartley Soulé who saved Bob Gilruth from boredom and introduced him to the world of aircraft flight testing in 1938. Now Soulé was in Langley's Tracking Unit and found himself working under the fellow whose career he had inadvertently turned into a quest. By late 1959, they'd settled on eighteen sites. Some were on military bases and already had radar units that could track the Mercury capsule. Others had to be built from scratch. Two would be aboard ships, one in the Atlantic Ocean, the other in the Indian Ocean.

When they ran the numbers on line-of-site transmissions from each station to the capsule, they decided that the site planned for Sydney, Australia, could be dropped. Australia had two sites anyway, at Muchea and Woomera. Canton Island and Kauai, Hawaii, adequately covered the rest of the Pacific Ocean.

Across the lower half of the United States, sites were planned at Point Arguello, California; White Sands, New Mexico; Corpus Christi, Texas; Eglin Air Force Base, Florida; and the launch site at Cape Canaveral. The British readily agreed to allow a site on Bermuda. Grand Bahama and Grand Turk islands also said yes. Grand Canary Island got a station to cover much of the Atlantic and northwestern Africa. Two more stations would be built in Nigeria and Zanzibar to complete African coverage and extend tracking into the Indian Ocean. Finally a reluctant Mexican government agreed to allow a site to be built at Guaymas, on the Sea of Cortés.

The worldwide tracking network wasn't nearly as exciting as the astronauts-in-training, or as the rockets and capsules to vault them into space. But it was just as important. On some missions, it would save lives.

The Chief Designer kept the pressure on from the beginning of 1959 to the end. He'd sent a probe close to the moon and into solar orbit. Then the Americans matched him, or almost, when Pioneer 4 finally broke free of Earth's gravity in March. But it missed the moon by even more than Korolev's machine, 37,300 miles.

Clearly both countries had work to do on space navigation. It wasn't a problem of knowing where to go. That much was obvious, and the laws of physics dictated how you got there; the route was always a curve through space. The problem was finding ways to precisely control the little thrusters that pointed a payload in the right direction and the big rockets that got the payload to the point where the little thrusters mattered anyway. If a rocket fired too long, or stopped firing too soon, or developed too much or too little power, not much else mattered. The payload wasn't going to get to its destination.

While both Russian and American engineers wrestled with those problems, astronauts were being named and became instant heroes. Korolev broached Nikita Khrushchev with the question of hiring cosmonauts and got a tentative nod to begin planning. Korolev's success in the space race, now being run in the open, might have had something to do with this approval. As he racked up successes, and gave the Soviet Union more and more propaganda victories, it grew harder and harder to deny Korolev whatever he wanted.

At the same time, versions of Korolev's R-7 rockets rolled out of assembly plants for delivery to the Soviet military. One by one, and only slowly one by one, they were fitted with atomic warheads and deployed as the frontline response to attack by the United States. The rockets diverted to the space program didn't always reach orbit when the Chief Designer wanted, but they were good enough to reach the enemy. (The R-7 was identified by the letter *A* by NASA and U.S. intelligence services. Korolev's people liked to call it the R-7M—"M" for modified. Depending on its mission, they would stack a third and even a fourth stage atop the basic rocket.)

Korolev quietly lost another moon probe in June. But on September 12, his R-7 rocket worked perfectly. Three days later Lunik 2 impacted on the moon. For the first time, an object made by humans had reached another planetary body. Nikita Khrushchev was in Hollywood that day, lecturing starlets on the set of the movie *Can-Can* about their skimpy clothing. His American tour made daily headlines, particularly when President Eisenhower denied him a trip to Disneyland on the ground that providing security in crowds of tourists would be impossible. The Lunik 2 story gave Khrushchev powerful ammunition as he claimed superiority for Soviet technology and by extension for the Soviet system of government. The next day he was at a farm in Iowa crowing about agriculture back home.

The space race gave him credibility. The much ballyhooed missile gap gave him the appearance of strength. Nikita Khrushchev was born a peasant. But he understood showmanship, misdirection, and the value of intimidation. He knew that strength included both what you really had and what your enemy thought you had. Nine months earlier, the number one hit on the American Top 20 was Conway Twitty's "It's Only Make Believe." It could have been Khrushchev's theme song.

After the Lunik 2 success, the Chief Designer decided that his cosmonauts did not need to be test pilots. The spacecraft being designed in his bureau was largely automatic; its passenger would have no control over any part of its flight. A regular jet pilot in fine physical condition could do the job. The word passed quietly throughout the Soviet Air Force that volunteers for a trip into space would soon be considered.

Then Korolev pulled off another stunner. Lunik 3 went on its way October 4, exactly two years after Sputnik 1. This time the goal was not to hit the moon, but to loop behind it with a camera. The spacecraft was the most complex yet built by either the Soviet Union or the United States. Korolev's engineers had developed and tested in-flight control thrusters that allowed the spacecraft to be pointed accurately. It was fitted with solar cells to provide electrical power. But that wasn't the end of it.

His trajectory experts plotted a course that put Lunik 3 into an orbit so stretched out that it was barely elliptical. Its high point carried it 56,000 miles beyond the moon. There it would begin to fall back toward Earth, looping about 25,000 miles at perigee and beginning to climb again. If its launch from the space base at Tyuratam in the Siberian steppes was timed exactly right, it would pass close to the moon on its way to apogee. Then as it climbed farther and farther away, it could look back and see what no one had ever seen.

Korolev harangued his team to get both spacecraft and rocket ready for the fortunate confluence of both the moon's proper position and Sputnik's anniversary falling on October 4. It worked. Then trouble: Lunik's cooling system began acting up. If vital instruments overheated, they would fail and so would the mission. But this spacecraft was built to respond to new commands from Earth, one of the first to have such capability.

While Korolev challenged the control center team, sometimes with beguiling words, sometimes harshly, they found an answer. Commands went to Lunik 3 and it began to rotate. Years later the Americans would call this maneuver the "barbecue mode." By turning slowly, a spacecraft absorbed solar heat, then radiated much of it away as it revolved. The technicians also commanded heat-generating systems to turn off until they were needed. The heating problem was solved.

The Tass news service announced the Lunik 3 launch but was cryptic about its mission—"a scientific payload." Tass reporters knew the whole story. Korolev had learned to include journalists among his captive audiences. The reporters were no dummies; when something went wrong, they filed their notes in wastebaskets instead of filing stories for international distribution. This time an overeager science correspondent talked too much. Yevgeny Bogoyavlensky mentioned Lunik 3's camera in a radio interview. "One can reasonably suppose that these unique photographs will be successfully transmitted by automatic radio apparatus when the station returns toward the Earth," he said.

When his comment was relayed to London, Tass quickly disavowed him, even saying that no such broadcast had occurred. A few hours later, its editors confronted with recordings, Tass conceded that Bogoyavlensky had been interviewed on the air, but said the camera comment was "his own supposition." If the camera system failed, the escape hatch was now re-opened for Tass to ignore it completely. The confusion over Lunik 3's mission made it an inside story for U.S. newspapers. The *Los Angeles Times* had a far better space story to play at the top of page one:

Future War on Moon Over
Lunar Rocket Stations Seen

Space-aviation editor Marvin Miles (the title was newly minted; Miles had covered aviation for years) quoted a Douglas Aircraft engineer speculating on a moon war. The moon was a perfect place for attack missiles to be based, R. L. Johnson said, because a moon base could see every country on Earth.

"America, with a philosophy of never firing the first shot, requires a certainty of retaliation rather than instantaneous retaliation," he said in the Miles story. "The ability to hit Earth targets from the moon—with plenty of time to launch and even escape—would be a tremendous deterrent."

But that could lead to a moon war as the United States and the Soviet Union fought over lunar territory, he said. Neither American nor Russian spokesmen commented on the story. It was the threat *du jour* and quickly passed from public attention. By then, Lunik 3 had passed the moon out-bound, and for forty minutes its camera, which did indeed exist, photo-graphed almost 70 percent of the lunar dark side.

The technological tricks weren't finished yet. For its time, the Soviet craft was a state-of-the-art one-hour photo shop. An automated system pulled the black-and-white film from the camera and developed it on board. Then a low-quality television camera focused on the processed film, began to slowly scan images, and transmitted them home to Mother Russia. At the same time, Earth stations receiving signals from Lunik 3—including England's famed Jodrell Bank radio telescopes—noted something strange. Instead of the steady beeps from the scientific station, they changed to a warbling sound. Sir Bernard Lovell deduced that photos were indeed being transmit-ted, much like the news services of the world transmitted wire photos. He set his equipment at Jodrell to decode the images and almost succeeded. But he guessed wrong on the scale; when he released a photo the next day, the moon's far-side craters were badly distorted.

The Soviet science team didn't have that problem. They were angry with Lovell for upstaging them, but quietly happy that he'd gotten it wrong enough to make their pictures look good in comparison. It sounds easy in the light of the technology that followed. For 1959, it was an amazing feat. Color television was years in the future. Some parts of the United States had only recently gotten TV stations, and most of the Soviet Union had none at all. Television cameras in studios were huge, mounted on dollies that required a man's strength to move about. Yet for this single purpose, Soviet engineers had shrunk a camera to fit inside a spacecraft. Picture quality was poor, but the pictures were good enough. They caught the world's imagination and were a true scientific accomplishment.

The moon's other side was a strangely unfamiliar place. It was far more heavily cratered than the face it showed to Earth, the difference between an acne-blotched teenager and the face he'd wear as a mature adult. One huge crater dominated the terrain. It would be named Tsiolkovsky for the Russian who had calculated space trajectories seven decades earlier. Scientists the world over hailed the photos and began to speculate on the two lunar sides. Theories eventually ranged from the front face being covered with a mile-deep layer of dust to the back side collecting most of the inbound meteors and debris that created the craters. Those arguments wouldn't be settled until man had walked there himself and brought home chunks to be studied in laboratories.

The Russians couldn't resist the opportunity to tweak the American nose one more time. They made a show of presenting far-side photos to the president of the American Rocket Society at its annual meeting in Washington. Of course Korolev wasn't there. The presenter was Academician Leonid I. Sedov, who only a few years earlier had left journalists scoffing in Copenhagen when he said that Russia would soon conquer space. While Korolev ran the show, Sedov took the bows. This time nobody scoffed.

Khrushchev couldn't fail to note the attention America's Original Seven astronauts got in the press. As he left the United States, an announcement came out of the Soviet Union that it, too, had cosmonauts in training. A few weeks later *Newsweek* ran a story with photos of a so-called cosmonaut named Aleksei Gracher. The short story intimated that Gracher and two others, Aleksei Belokonev and Ivan Kachur, were training for a space mission. The story was phony, probably a plant by the Soviets. Korolev's man-in-space program wouldn't pick its first cosmonaut trainees until the following spring. Who were Belokonev, Gracher, and Kachur? The names did not appear again. The Soviet premier might have hummed "It's Only Make Believe" all the way home.

The search for cosmonauts was, however, beginning. When word of Lunik 3's photos spread across the Soviet Union, one young man was particularly inspired with his country's new space program. He was small for an Air Force fighter pilot, needing a pillow to raise him up enough to see out of some of the planes he flew. But he was smart and courageous, and on the day Lunik 3's pictures were radioed home, he made a decision. They were asking for space pilots and he wanted in.

His application went forward the next day. The name on the top line meant nothing to the Chief Designer or to the doctors who would be critical to the selection process.

He was just another of many brave young pilots who volunteered, this Yuri Gagarin.

It had been planned for months, so it was just a coincidence that America's seven astronauts were on the West Coast and highly visible when the Lunik 3 story broke. It wasn't Bob Gilruth's intention to divert attention from the Russians. But if it worked that way, he wasn't going to complain. He had some help; in Los Angeles, the biggest story was the World Series between the Dodgers and the White Sox.

It seemed like a good time to send the astronauts on a tour of the burgeoning aerospace industry in Southern California, while letting them be highly visible at the Society of Experimental Test Pilots convention. SETP was only a few years old in 1959. Its founding members were about fifty of the nation's best pilots, and more than half of them would die on the job. Those who didn't would sooner or later miss being killed by only a hair. People who went to work in the morning as test pilots knew that getting through the day took a little luck and a whole lot of skill. On other days, it took nothing more than the alarm clock going off on time.

As it turned out, a test flight of the X-15 rocket plane was on the schedule in early October at Edwards Air Force Base, about sixty miles northeast of L.A.

The NASA group stayed overnight at the Antelope Valley Inn, familiar territory for many of them, and drank a bit with old friends. At dawn they gathered in the chilly desert to watch the flight, standing in small groups on a knoll that gave a perfect 360-degree view and drinking badly needed coffee poured from steaming Thermos bottles. Nearby a sedan with old horn-type speakers on its roof kept them in the know. Walt Williams was there, along with one of Bob Gilruth's assistants, Kenny Kleinknecht. Both were in radio contact with base operations and with the B-52 rolling down a runway a few miles away. Williams was nearing the end of his tour as NACA's—now NASA's—operations manager overseeing high-performance X-craft and other research aircraft. He'd been in the desert more than twelve years.

In the new scheme of things at NASA, airplanes were rapidly taking a backseat to spaceflight. Bob Gilruth already had told Williams that he'd be needed on Project Mercury. Nobody had his level of hands-on operational experience.

The X-15 was tucked up into the B-52's belly, and with Scott Crossfield at its controls, this was to be its first powered flight. Climbing slowly, the big modified World War II bomber scratched toward 26,000 feet, where the rocket ship would drop away and then blast upward for just forty seconds, going high, but not yet to the fringe of outer space at 200,000 feet and more. Those flights would come as step by step, the X-15's capabilities were explored. With the bomber beginning to leave white contrails in the blue sky above, the group saw a dust trail coming their way across the desert floor. A fast-moving station wagon soon appeared.

It screeched to a stop and an apprehensive young man hopped out. Walt Williams immediately charged him, stopping inches from his nose. "You're two hours late," he screamed. "Where the hell have you been?"

"I overslept . . ."

"If you have that problem, you just don't go to sleep. You want to stay in this job, you *never* oversleep. You're supposed to be briefing those guys over there about the X-15."

"I . . ."

"You ever do this again and you'll never fly another flight, if I have anything to say about it."

"I . . ."

"Now get over there and do your job. They're about to drop the X-15. Go! Get!"

"Yes, sir."

Paul Haney of the NASA public affairs staff heard it all as he poured more coffee and wondered who the young guy was. He found out a minute later. Chastised and downcast, Neil Armstrong trotted over to the seven astronauts and their NASA support staff to begin his briefing.

The X-15 flight that day was perfect. So was Armstrong's on-time performance for the rest of his test pilot career.

Gus Grissom was due in San Diego. Paul Haney went along, to handle the inevitable press inquiries and to be the buffer between the astronaut and reporters. Grissom worried him a bit. The shortest of the astronauts was lively and ebullient among his friends, but he was the most taciturn of them all in public. Now he was scheduled to give a pep talk to the workers at General Dynamics.

The rocket builders had good reason to be excited. Only four weeks earlier the U.S. Air Force declared Atlas operational as an intercontinental ballistic missile. Two of the rockets had been fired since then, on 5,500-mile flights from Cape Canaveral to the South Atlantic Ocean, and a third already was on the pad. On each of these quick-succession launches, the targeting and aiming software got better.

Democrats in Congress and contenders for the White House in the next year's election still harped about the missile gap, and the press contributed to fears that a Soviet first strike was a distinct threat. Nikita Khrushchev knew better, but he wasn't going to admit that his rocket man, Sergei Korolev, was running on a thin budget and still suffered more failures than successes.

So far, NASA planned to use the weapon Atlas, unmodified, as its primary Mercury rocket. So General Dynamics had a double-header winner on its hands, with orders for new rockets stacking up. Closing the nonexistent missile gap was a national priority, and so was putting Mercury astronauts into space.

When Haney and Grissom arrived in San Diego, General Dynamics had virtually shut down its plant, and eighteen thousand men and women were waiting outside in high excitement to see this astronaut and to hear him speak to *them*. Grissom was introduced and the crowd erupted into a long and sustained roar of approval. It went on and on. Finally Grissom held up his hands, and gradually the noise receded until there was complete silence.

Haney had no idea what Grissom planned to say. He'd declined the offer to have the public affairs guys write a speech, and didn't seem to have one of his own prepared. It looked like this one would be straight off the cuff.

Grissom stood at the microphone, looking out at people who might hold his life in their hands. If the day came when he strapped his butt into a

Mercury capsule and an Atlas rocket launched him toward space, it would be these people who built that rocket. *Low bidder or not, you guys better do it right!*

What other thoughts flashed through Grissom's mind at that intense moment were never shared. He stood there staring at them, finally took a deep breath, and gave his speech.

"Do good work," he said, and sat down.

The silence was complete. Then it sank in. This little astronaut with the crew-cut hair and uncomfortable-looking civilian suit had said it all. *Do good work! Damn right we'll do good work!* The roar of eighteen thousand voices screaming their approval and their promise shook the courtyard walls and nearly knocked Grissom and the others off the makeshift stage.

Haney forced himself not to laugh. "Do good work!" he said later. "A three-word speech! Hell, maybe all speeches should be that short."

Chris Kraft was the first speaker at the SETP convention. Slender, dark-haired, and five-ten, he was only thirty-five and could have passed for one of the seven. Many of the test pilots in the Beverly Hilton meeting room knew him personally, from his fifteen years in flight test operations at Langley. He was a cool character, unflappable in emergencies and with a reputation for making snap decisions that were right on. He and Deke Slayton were a two-act team, sent here by Bob Gilruth to give one of the early public presentations on exactly how Mercury would get men into space. For a program only a year old, an amazing amount of work had been done. Kraft launched into his lengthy speech. He was precise and technical, describing how launch, orbital entry, and retrofire would work—crucially making the point that the astronauts would have some control.

That marked a critical difference between the philosophies of Bob Gilruth and Sergei Korolev, though nobody outside of the Chief Designer's inner circle knew it in October 1959. Gilruth grew up in his profession side by side with pilots. He understood how important they were, and even when he ran the Pilotless Aircraft Research Division, he never forgot that the ultimate goal was to put a man in control of the machine. Korolev knew pilots too, and was one, but his faith in the machine superseded his faith in the man aboard. When the space race was in its middle laps, Korolev switched sides. A man in control of his spaceship would become mandatory.

The test pilots ate up the details provided by Chris Kraft; Marv Miles was bored. His story in the next day's *L.A. Times* didn't even mention Kraft.

Then Air Force Captain Donald K. Slayton, "Deke" to his friends and eventually to almost everybody, took over, and Marv Miles had him on the front page of the next day's *Times*. Deke's speech was fourteen pages long. He was talking to his own, and among them were men who would become astronauts in years ahead. Even those like Chuck Yeager, who disparaged Mercury as beneath the dignity of real pilots, wanted to know the inside story. Deke got everybody's attention with his opening sentences.

"I will discuss the program from the inside out, or pilot's point of view," he said. "In accordance with one publication's personality analysis of the astronauts, we should have the poised and articulate faction here today for optimum performance of *this* mission. However, some facets of the discussion may lend themselves almost equally well to the stubborn, frank approach which is theoretically my forte."

The room rippled with chuckles. Slayton was purely their own. If he lacked the elegance of John Glenn's podium presence, the wit of Wally Schirra, or the articulate schmoozing of Al Shepard, he was the guy who would always tell it like he saw it. And they'd all heard about Gus Grissom's speech at General Dynamics. Slayton already had talked a full thirty seconds longer than the littlest astronaut; he was barely getting warmed up and he knew his audience.

"First, I would like to establish the requirement for the pilot, or astronaut, in Project Mercury and follow-on programs," Slayton said with a sardonic grin. "This may seem a ridiculous and unnecessary approach to some of you since the object is to put a man into space and determine what his capabilities are.

"However, exceptions or objections to the pilot range from the engineer, who *semi*seriously notes that all problems of Mercury would be tremendously simplified if we didn't have to worry about the bloody astronauts, to the military man who wonders whether a college-trained chimpanzee or the village idiot might not do as well in space as an experienced test pilot."

Now they were roaring with laughter. Slayton was touching on some home truths that all of them had heard before. It was time to get serious.

"The answer to the engineer is obvious and simple," Slayton said. "If you eliminate the astronaut, you concede man has no place in space."

None of this made the newspapers. Any dissension in the NASA ranks was purely an internal affair, and reporters had yet to cultivate sources who would leak such details. So Slayton got in his digs and then turned serious. For the first time, the world would hear an astronaut discussing the Mercury capsule and missions.

He talked about controls. Astronauts would not have control over the Atlas rocket, but once in orbit they'd control the capsule itself. On the way up, they could activate the escape tower mechanism if things went terribly wrong. When it was time to come down, the astronaut had override capabilities to set up the capsule for retrofire and to fire the rocket pack that would get him out of orbit and on the way home.

He told them in detail about the instrument panels, the electrical circuits, the warning lights that would turn from green to red just like in the airplanes they all flew. He told them about the parachutes that would lower the capsule into the ocean at the end of reentry. No runways for this machine.

Mercury couldn't do much more. There was no rocket pack to raise or lower the orbit; once the Atlas had done its job, you got what you got until it was time to come down. There were no thrusters that could move the capsule sideways in orbit either. An astronaut was essentially on an elliptical roller coaster. He could turn the capsule in any direction, but he couldn't move it off that invisible orbital track. There were looks of consternation from the pilot-filled audience.

"This is not compatible with aircraft practice and should be designed out of future space vehicles," Slayton assured them. "We don't expect this first vehicle to be the ultimate."

The big news was Slayton's comments on the missions themselves. The capsule was designed to stay up there for eighteen turns around the Earth, about twenty-seven hours in all. An astronaut would eat and sleep in orbit. Early flights would be unmanned and automated, he said, followed by an astronaut going up for a four-and-a-half-hour, three-orbit trip. The long flights would follow. It was heady stuff, but Slayton was too honest to ignore the obvious.

". . . if everything works perfectly the pilot's task will be quite simple," he said. "He can devote the entire time of flight to observing and analyzing his and the vehicle's reaction to space travel and should return with an idea of man's capabilities in a space environment." And if an emergency cropped up, or equipment didn't work right, he added slowly, "we would have an even better idea of man's abilities in space." Like all the test pilots in the room, Deke Slayton was a realist. And a darned good prognosticator.

That night with his new astronaut buddies and his old test pilot friends, Slayton got rip-roaring drunk. Wally Schirra saw his chance for a gotcha. With Deke passed out, they loaded him onto a cot and carried him through a second-floor access door onto the top of the Hilton marquee. Maybe it was the morning traffic on Wilshire Boulevard. Maybe it was the bright sun filtering through the palm trees. Maybe it was just time to crawl up through the hangover and return to consciousness.

When Deke pried his eyes open, the glare and the noise hit him like a hammer. *Where the hell . . . ?* He opened one eye and squinted at the world until he figured it out. *Okay, guys. The next gotcha is mine.*

When Dr. Randy Lovelace joked that he hoped these astronauts would never give him a physical, he put his finger on the crack that turned into a major schism between the astronauts and the flight surgeons. Before it was over, the argument handed the Russians their biggest space race victory since Sputnik. That bitter day wouldn't come until 1961. But its effects linger, and at least one high former NASA official still looks back from the threshold of the twenty-first century and sneers about the "so-called medical community."

There was nothing like that kind of animosity on the engineering side. Early on, Bob Gilruth called the astronauts to his office and told them his ground rules. "We picked you fellows because you're test pilots," he said. "We have engineers and they're already working on the Mercury spacecraft. But anytime you see anything, anytime you think something needs more testing, or even should be redesigned, I want you to let me know."

Gilruth's rule was cooperation. He meant it, and the astronauts took him up on it. One of the first things they saw was that the Faget-Johnson-Mathews design for the capsule had no windows. The astronaut inside would be Spam-in-a-can. There was a periscope, but its field of view was limited and it was awkward to use. If they were to observe the Earth below in any detail, and if they were to look at stars and planets for scientific purposes, they needed a window. Gilruth listened, turned to Max Faget, and told him to make it happen. Putting a window into a capsule sealed against an outside vacuum required major redesign and even structural changes. Faget didn't argue. It was the right thing to do.

Arguing with the doctors was never so easy.

The seven hardly had time to put their affairs in order and make the move to Space Task Group headquarters at Langley Air Force Base, Virginia, before the so-called medical community started getting in its licks. "The doctors were concerned about a lot of weird things," John Glenn remembered. "Some of the things they put us through were pretty extreme."

Nobody knew in 1959 how a man's body would react to either zero gravity or the high gravity forces inflicted during liftoff, unspecified in-flight emergencies, reentry into the atmosphere, and the various abort and malfunction events that could occur on liftoff or landing. The theories started with complete malfunction of almost every organ and muscle. Military flight surgeons have the final word on whether a pilot is fit to fly or not. They carried that attitude into the Mercury program. They started with their disaster scenario, and their attitude was simple and firm: "Prove us wrong."

The astronauts and other test pilots couldn't shake them. Bob Gilruth tried, but didn't have the authority to overrule them. The underlying threat was always there. If something bad happened, there would be an investigation. If the investigation showed that the flight surgeons had given due warning and been ignored, heads would roll. They gave their due warning early. The job was to prove them wrong.

An aircraft flying a high parabola could get maybe thirty seconds of zero gravity as it went over the top and started down. Once the X-15 rocket plane went fully operational, its pilot might experience four or five minutes without gravity. Neither came close to being enough time to tell the medics that the human body would or would not function during hours or days of spaceflight.

The only way to get that data was to go into space. So the monkey flights were conceived. Some number of test flights, suborbital and orbital both, would be needed to prove out the Mercury capsule and its rocket. If astronauts couldn't be used as test animals on some of those missions, monkeys and chimpanzees could. The medics would get their data. That covered most of the bases. But a monkey was not a man. The doctors wanted man data. They couldn't get the zero g, but they could get the other extreme. The U.S. Navy had a man-rated centrifuge at Johnsville, Maryland. They'd program it for the launch and reentry g-profiles, and send the real thing—Mercury astronauts—to ride it.

Those rides were a piece of cake. During launch on an Atlas rocket, engineers calculated that astronauts would experience 7 or 8 g's and that reentry g-loads would peak between 10 and 12. Neither peak would last more than a minute or two, and a human body enduring 12 g's still functioned. It took considerably more before a man passed out.

To make things easier on the astronauts, Max Faget's group designed a padded form-fitting couch and tested it with human subjects at Johnsville. It made the g-loads much more bearable. Each of the seven rode launch and reentry profiles on the Johnsville centrifuge, and many times. Couches were molded to fit each astronaut's body, with padding in the right places for maximum comfort.

It was the other centrifuge tests that hurt. The doctors concocted a whole series of them. They started by running Shepard, Cooper, and Glenn up to 16 g's. "That's a real gut-buster," Glenn remembers. "You even think about relaxing and every corpuscle wants to give up." But they passed with flying colors, including the red faces and black-and-blue bruises those runs caused.

Then the medics looked at Mercury's contingency plans. In a normal mission, the capsule would descend to an ocean landing under its parachutes. But what if it had to come down on land? NASA was in the early stages of developing its what-if scenarios designed to ensure safety out to three nines—the odds of success had to be .999 on the plus side. Engineers learned to think about everything imaginable that could go wrong. Inevitably things not thought of would go wrong. But never twice.

The worst-case scenario for a Mercury capsule coming down in a desert included high winds that sent it tumbling end over end. If that happened, the astronaut inside would be thrown about, subjected to high g-forces, and probably injured. Their answer was to look at redesigning the shoulder and seat belt harnesses. But before that was ordered, they put Shepard and Glenn through "tumble runs" on the centrifuge. "We called them e-i, e-o," Glenn said, "eyeballs-in, eyeballs-out."

The pod at the end of the centrifuge arm could be rotated in any direction. Tumble runs rotated the pod every two seconds, throwing the astronaut inside from positive to negative g's with each rotation. As the centrifuge spun up, they went from two by two, to three by three, and finally to four by four. "That's a delta of four positive to four negative g's in two seconds," Glenn said. "And that really popped you out. You could see the pattern of the shoulder straps in petechiae [tiny burst blood vessels] on your body." One of the flight surgeons insisted on doing a run himself. They ran him to five by five and he could barely breathe. "They figured that his heart was swinging out and behind, and knocking the air out of one lung," Glenn said with a shake of his head. "Al and I terminated after the four by fours."

Another concern involved a capsule spinning out of control in orbit, perhaps from a stuck thruster. At what point did the astronaut phase out and not be able to function? A capsule was rigged into a setup with triple gimbals and six degrees of freedom. It could spin, twist, and turn in any direction, including several at once. The capsule was fitted with air jets and the astronaut inside had his hand controller to fire them and bring the tumbling craft under control. They started with tumbling in one axis only, bringing it up to 30 rpm. Neither Glenn nor Shepard had any trouble getting it under control. So they tried two axes. Same results. Finally they went to a full three-axis tumble. Even at 30 rpm, neither man failed to stop the rotation. "We were going tail over teakettle," Glenn said, "but we proved that man could function in the worst-case scenario."

Finally the medics admitted that these particular supermen could handle almost anything involving high g-forces. They reserved judgment on zero gravity until the monkey flights started.

It was a monkey flight that made Al Shepard mad and gave America another black eye in the space race. But that was more than a year away. And it wasn't a flight surgeon who threw the knockdown punch.

5

Moscow hosted a major exposition in the spring of 1959. The U.S. exhibit included a typical American kitchen supplied by the Westinghouse Corporation. Vice President Richard Nixon was there for the unveiling, and so was Soviet Premier Nikita Khrushchev. When Nixon showed the premier through the U.S. exhibit, he extolled the virtues of American society. Khrushchev listened for only a few moments before challenging him in front of reporters and cameras. The so-called Kitchen Debate turned into a loud, finger-wagging confrontation over the merits of socialism versus American-style democracy. Then a funny thing happened.

When the verbal imbroglio ended, reporters and photographers wandered away. Nixon took Khrushchev behind the exhibit's panels and kitchen equipment, where they had a polite and respectful conversation. Nixon was a big supporter of the space program inside the Eisenhower administration and was intrigued by the Soviet Union's smashing successes over the previous eight months. That night Khrushchev related the story to his family over dinner in the Kremlin.

"This Nixon is a funny fellow," he said. "The first question he asked was what kind of fuel we used in our space rockets. I told him I don't know. I have people who handle that sort of thing." Then he chuckled and bobbed his head. "You know, maybe that Nixon is really with the CIA."

They'd given the contract to build the Mercury capsule to the McDonnell Aircraft Corporation in St. Louis in January 1959. By midyear, they were testing pieces of it. Sometimes the tests worked.

Max Faget's Little Joe booster, a short and squat rocket intended only to get space capsules high enough to test them without putting them into orbit, was ready in July 1959. Its first mission off a pad at Wallops Island, Virginia, was a laugher. It went exactly nowhere. Little Joe 1 was supposed to send a dummy Mercury capsule topped with a live escape tower out over the Atlantic. When the stresses of acceleration and aerodynamics hit maximum, the tower was supposed to fire and pull the capsule safely away. After a month of technical and weather delays, they were thirty-five minutes from launch when the escape tower decided to go it alone. Surprised controllers and technicians heard an awful sputtering roar and turned to see a trail of smoke and the Mercury capsule being lifted out over the sea. The Little Joe stayed sedately behind.

At about two thousand feet, the escape tower let loose of the capsule, darted off to the side under power of another small jettison rocket, and died. The capsule dropped into the water a half mile offshore. All they learned was that the Mercury escape tower had enough power to do an off-the-pad abort. But that wasn't the plan on this test. It was marked down as a failure and the problem traced to an electrical coil that overloaded and sent the "fire" signal to the tower.

A few weeks later they tried a Big Joe test from Cape Canaveral. Big Joe wasn't a rocket, but the code name for launching a Mercury test capsule atop an Atlas rocket. This was a reentry experiment to try out the Mercury heat shield. Working with engineers at McDonnell Aircraft, Max Faget and the Space Task Group designers had settled on a secret ablative material for the shield. Ablative meant that the fiberglass-like shield shed heat by charring. Little pieces would flake away, but the more it charred, the more impervious to heat it got. That was the theory, anyway, and it worked in the laboratory. Whether it worked when it slammed into the atmosphere at fifteen thousand miles per hour or more was the question. With the fires of reentry reaching four thousand degrees, everyone assumed that astronauts would prefer that it worked. Cremation was not in the flight plan.

Big Joe was delayed too. First the Air Force had problems getting the Atlas rocket just right, then the NASA people ran into trouble with their instruments. Finally in the 3 a.m. black of September 9, it lifted off from an Air Force pad at Cape Canaveral.

There are three call-outs during an Atlas launch that would later become familiar to Americans listening to radios or watching television during Project Mercury. The first is "Liftoff."

The second comes a few minutes later. "BECO." *Bee-ko.* Booster engine cutoff. The big outboard rocket engines built by the Rocketdyne Company shut down and then drop away.

Now another rocket, the center-mounted "sustainer engine," keeps firing high in the atmosphere and pushes toward space. Or if the payload is an atomic warhead, the sustainer engine sends it on its long, looping trajectory toward a target.

The third call is "SECO." *See-ko.* Sustainer engine cutoff. If SECO comes at the right moment, it is the surest sign that all is well. There are other things that can be wrong, of course—the trajectory, for instance.

Two minutes after Big Joe left Florida behind, instruments in the blockhouse went haywire. Nobody called out "BECO," and if there's no BECO, there's not much chance for SECO.

Then it got strange. Signals from the Mercury capsule kept coming in and they looked pretty good. The odd thing was that, if the capsule had separated from the Atlas, it hadn't flipped around to present its heat shield BEF—blunt-end forward. It seemed to be ramming toward the ocean like a hypersonic bullet. Then the signals disappeared, lost in the ion sheath of reentry heat.

They figured it out later. First the computers calculated that the Atlas booster engines had stayed around for the entire ride. The extra weight chopped about eighteen hundred miles per hour from the final velocity and meant that if it survived at all, the Mercury capsule was coming down five hundred miles closer to Florida than planned. A Navy airplane picked up its radio beacon and found it just before dawn, floating peacefully in the center of a green-dyed patch of the Atlantic.

The Mercury people heard the news and were happy and confused. The capsule should have burned to a crisp. The tapes from its onboard instrument recorders, and from Air Force radars and computers, told a weird and wonderful story. Instead of heading for orbit, the programmed trajectory called for a turn-around with the sustainer briefly ramming the capsule into the atmosphere before turning it loose. When the booster engines didn't fall off, the sustainer engine kept trying. But it came back down sooner, and much steeper, than planned.

The little capsule's instruments saw what was happening and fired its own thrusters in a vain attempt to get everything back on course. Finally it ran out of fuel and cut itself free from the Atlas. It was going fifteen thousand miles an hour, was sixty-five miles high and out of fuel, and pointed down small-end first. The speed was higher than the test called for, and the temperatures were hotter. But the shape was right. When Faget, Caldwell Johnson, and Chuck Mathews sketched it out a year earlier, they designed it to flip around from nothing more than the forces of gravity and the atmosphere. So it did.

Now reentering in the approved BEF configuration, the capsule did its job. At about fourteen miles up, its drogue parachute popped. A little while later the main chutes deployed. It splashed down comfortably in the Atlantic, dropped a SOFAR bomb signaling device which exploded underwater and was heard by waiting ships, flipped out a package of sea-marker dye, deployed its radio antenna, and chirped a homing signal picked up quickly by rescue airplanes.

Late that night, the capsule came home. Technicians opened it and pulled out an envelope. It was addressed to Bob Gilruth. He took it with a puzzled twinkle in his eye, ripped it open, and read the message inside.

This note comes to you after being transported into space during the successful flight of the "Big Joe" capsule, the first full-scale flight operation associated with Project Mercury. The people who have worked on the project hereby send you greetings and congratulations.

Gilruth chuckled, shook hands all around, and declared the capsule's mission a success. The secret heat shield worked and the Mercury capsule was even better than they'd hoped.

The next day Gilruth wasn't chuckling. The Air Force had declared the Atlas launch a failure in a press conference and that's what made the news. It took hours for the recovery forces to find the Mercury capsule floating in the Atlantic, longer to bring it ashore. By the time Gilruth's people knew the real story, the afternoon newspapers, the radio and television news shows, and the next morning's papers all proclaimed the test another failure.

Walt Bonney put his public affairs people to work on a quick news release and on the phone calling reporters. It was no good. The revised story got little or no play. Gilruth was bitter and the young NASA learned an unfortunate lesson about news cycles. "The press doesn't like anything that's over a few hours old," Gilruth said later. "We didn't have it when they wanted it. By the time we had it, they didn't want to publish it—especially since it was a success."

So for the first time, a rocket failure led directly to a spacecraft success and nobody outside of NASA cared. There would be another failure down the road that put Gilruth and the Air Force into a fight that took them all the way to the inner ring of the Pentagon.

The monkey flights started in December 1959. So did an internal argument over mission control. Sam the rhesus monkey rode a rocket and became famous. Then Chris Kraft came close to punching out Wernher von Braun. Reporters didn't hear about that one.

The doctors had made so much noise about the possible dangers of zero gravity that Bob Gilruth gave in and approved a series of biomedical experiments involving primates. Chimpanzees had ridden the centrifuge at Johnsville, Maryland, for years, endured other high-stress tests conceived by the aviation medics, and generally did no better and no worse than humans. Gilruth reasoned that it couldn't hurt to let the doctors run some zero gravity tests. Hardly anyone with an engineering degree thought the doctors were right, but it would take hard facts to make them back off.

Little Joe 2 was ready at the Wallops Island launch pad on December 4. So was Sam, named for the School of Aviation Medicine at Brooks Air Force Base in San Antonio, where he'd been raised and readied. Sam even had his own custom-made astronaut couch, an exact miniature of the real thing. The solid rocket Little Joe's job was to get the Mercury capsule and its escape tower to about 100,000 feet. It would be a moderate kick in the butt for Sam, but nothing a rhesus monkey—or a human—couldn't handle. After separation, the escape tower would take the capsule to nearly 400,000 feet. As it arced over the top and began to fall toward reentry, Sam should get about four minutes of weightlessness.

He didn't have to do anything but survive. As the countdown reached zero, a couple of astronauts watched with pointed interest. Al Shepard and Gus Grissom were arguing that a Little Joe should carry one of them on a capsule test. That would make the doctors shut up.

Little Joe belched smoke and fire and rose perfectly into the sky. After it disappeared from sight, it wasn't so perfect. The hardware worked fine. The weather didn't cooperate. High-altitude winds slammed the capsule and escape rocket, slowing its climb until it finally topped out at only 280,000 feet. Sam-in-the-can felt like he was going over the top of a roller coaster, lifting up minutely in his couch and weighing exactly nothing. Three minutes later he pressed back into the couch when gravity returned and the capsule roasted down into the atmosphere.

Inside Mercury, Sam was just fine. The parachutes opened, he dropped harder than planned into the Atlantic, and then he waited. It took six hours for the recovery forces to find him—he'd landed well short of the planned spot—and by the time they got him out of there, Sam was not a happy monkey.

While the doctors analyzed Sam and the data from his Little Joe flight, Chris Kraft went to Dallas to present a paper on how NASA intended to keep track of and control space missions. The NACA crowd grew up in their profession with communications and control; it was second nature. The flight research people like Gilruth, Faget, Kraft, and all the rest prepared detailed plans for each test flight. The test pilot took the airplane up and followed the plan, or at least as best he could and according to his own on-the-spot judgment. From beginning to end, the pilot and the ground controllers were in radio contact. The test pilot described everything—instrument readings, vibrations, aircraft handling, how he felt—over the radio.

So why should spaceflight be any different? Full-time communications and control was impossible. That had already been established. Once an astronaut was in orbit, he would pass in and out of communications range. On the other hand, mission controllers would have full-time coverage during the launch and again during most of the reentry phase. So that was the plan Kraft presented in his paper.

A mission control center would be designed and built. Controllers would monitor things and stay in touch with the astronaut. They'd advise him when that was necessary and listen to his reports. In an emergency, mission control would work with him to find answers. And most important, mission control was the boss. When push came to shove, the astronaut would do as he was told.

Not everyone agreed. An astronaut couldn't fly the rocket. At best, he only needed instruments that told him whether it was working right or not. If things went wrong, he could push an abort button. He didn't need some technician in mission control miles away to make that decision for him. In space, it was the same. Mission control could be thousands of miles away. Sure, it was a good thing to stay in touch on the radio, at least as much as possible. But if something went wrong a hundred miles up over New Guinea, what could a technician in Florida do about it? One expert at Bell Labs said that a desk with a few phones was enough to cover the contingencies.

Kraft's position—and NASA's—was that well-trained people armed with instruments, computers, and a steady flow of data could do plenty. He made his point at the symposium. Spaceflight was a new adventure, literally a new frontier, and it demanded more ground control, not less. And if something went cataclysmically, fatally wrong during a mission, Kraft and all the rest of them wanted every scrap of information they could get to find out why.

At a cocktail party that night, Chris Kraft ran into the other opinion. Wernher von Braun thought Kraft was wrong. He was probably feeling good with his own newly gained power. NASA headquarters had just won a fight with the Army over the fate of the Army Ballistic Missile Agency and von Braun himself. Much of the Army's physical plant and many of its people at the Redstone Arsenal in Huntsville, Alabama, were about to be transferred to NASA. The Army didn't give up easily. NASA counsel Paul Dembling negotiated the deal, quickly understanding that the Army's big concern was that the air force would somehow, someday, take over NASA. They settled on a ninety-nine-year lease for the physical property, with a clause reverting ownership to the Army if NASA left or lost its independence. "That way the Air Force would never get the property," Dembling said.

The new center, charged with developing rockets for NASA, was to be named the George C. Marshall Space Flight Center. Its director had just been named: Wernher von Braun.

So when von Braun challenged Chris Kraft at a cocktail party over his extravagant ideas for a mission control center, it was not a battle of titans. Kraft was merely a deputy director of a division inside a center. Von Braun was top boss of a center, the equal of Bob Gilruth.

The German contingent coming into NASA was not particularly popular. Gilruth himself felt animosity, left over from his active involvement in perfecting aircraft to fight the Nazis during World War II. Kraft agreed, but added a natural disdain for icons to his personal feelings. Through the fifties, von Braun got more than his share of publicity. He was handsome, charismatic, and had his own public relations man, a former newspaperman named Bart Slattery. Kraft didn't see von Braun as a team player, and now he was going to be part of NASA.

And now von Braun was picking holes in Kraft's paper about mission control. The German flew airplanes himself and agreed with the pilot-in-command philosophy. The man in control of the ship should make the decisions.

Kraft disagreed. As the circle of onlookers around them gradually grew, the discussion became more intense. The German began to raise his voice. This no-name from Langley was daring to challenge a center director. If Kraft thought about their relative positions in the new hierarchy, it was briefly. He was not a man to be cowed, particularly when he thought he was right. His voice got louder too.

Neither of them noticed that the hotel suite had become silent. Every eye was on the two, watching and listening with utter fascination. Von Braun was adamant. This mission control fixation of Kraft's was a bad thing. Kraft felt himself getting angry. *The logic is on my side. And the decision is already made anyway. Why can't this stubborn Hun admit it?*

In another minute, said Kraft years later, he might have poked von Braun in the nose. He didn't get the chance. Maria von Braun stepped between them, took her husband's arm, and whispered something in his ear. He didn't click his heels or utter a parting oath. What Wernher von Braun did was to allow his wife to lead him away and out into the hallway. Kraft took a deep breath, conversations resumed in the suite, and he walked over to the bar to pour himself a stiff Scotch.

That was just the beginning. Tensions between Gilruth's Space Task Group and von Braun's Marshall Space Flight Center would get a lot worse.

6

With the dawn of 1960, the race was into its fifth year. It had gone from that secret Soviet decision in August 1955 to launch a satellite to full-fledged competition with the United States for control of the high frontier and for an incredible propaganda victory.

Ask anybody on that New Year's Day and you'd hear that the Soviet Union was winning in space, that the missile gap left Western democracies dangerously exposed to a preemptive strike by that brute Khrushchev, and that the contest between socialism and democracy was still a toss-up.

The last two conclusions stemmed directly from the first. The Soviet space program had a flair for the dramatic. Its successes were spectacular and its failures were unknown to the public and only suspected by the world's intelligence services. Nobody seemed to realize that the United States, off to a slow start in the space race, was rapidly catching up. Even ignoring the number of Soviet failures, the launch numbers told that story.

1957:
Soviet Union, 2 satellites, 0 failures
U.S., 0 satellites, 1 failure
1958:
Soviet Union, 1 satellite, 4 failures
U.S., 7 satellites, 10 failures
1959:
Soviet Union, 3 satellites, 1 failure
U.S., 11 satellites, 8 failures

The Soviet failures were not announced. The illusion of perfection yielded an illusion of superiority.

Of the Chief Designer's four failures in 1958, three were moon probes and all came after the second Sputnik 3 reached Earth orbit. His four launches in 1959, three of them successful, all aimed at the moon. Sergei Korolev was a single-minded individual. Once he mastered the art of getting into Earth orbit, he abandoned its potential and aimed for the moon.

The U.S. space program diversified from the beginning. Its seven successes in 1958 included two lunar probes which sailed past the moon without hitting it, four scientific satellites in Earth orbit which made major discoveries including the Van Allen Belts, and the first-ever communications satellite, Score, which transmitted taped messages back to Earth for thirteen days. The eleven American successes in 1959 included six Discoverer satellites that set the stage for military reconnaissance from space. The other five were a lunar probe that again missed the moon and four scientific satellites that relayed important data on solar flares, radiation, and micrometeorites. One of them took the first clear picture of land masses below.

In the propaganda war, the numbers game didn't matter. The Soviet space program had all the important firsts—first satellite, first dog in space, first probe into solar orbit, first probe to hit the moon, first pictures of the moon's far side—and a wily leader in the Kremlin who bluffed and blustered and took his country to the world's center stage.

Joe Average couldn't comprehend the scientific excitement over Van Allen Belts or solar flares, and in practical terms he had it right. But Mr. Average understood Nikita Khrushchev when he bragged about Soviet rockets and satellites and rattled the nuclear saber. Perception counted more than fact at that point in the space race. The next lap in the race boiled down to the question of men. Which country would put one into space first?

Perception also made the race a sternly serious business for both sides. Maybe too serious. Nobody laughed in public. The Original Seven had their gotchas, but they were private jokes. Not many people understood the value of humor to break tension and lighten up attitudes better than Wally Schirra. But there was one, and he was anything but private.

He was born November 23, 1959, live before a national television audience on *The Steve Allen Show*. It would be a few months before he became an astronaut and *persona*.

His name . . . Jose Jimenez.

Sergei Korolev shifted gears in 1960, and Bob Gilruth picked up the pace. The Chief Designer lost another moon probe, took a couple of vain shots at Mars late in the year, but turned most of his energy to prototypes of manned spacecraft. Once again he got away with hiding his failures. Gilruth was never that lucky, complaining that the press saw every failure, then hid some of his successes, or missed the point entirely.

Miss Sam got her ride on Little Joe 2 in late January and nobody could persuade reporters that the real purpose of the mission was to check out the hardware, not the rhesus monkey. She did have more to do than Sam. The medics had added a lever for Miss Sam to pull when a light flashed during the flight.

It was a more severe ride than Sam's too. This was a test of the escape tower system. The escape tower's solo act a few months earlier didn't count. This time, Little Joe got to about fifty thousand feet when the tower's rockets fired. Miss Sam endured a sudden spike in g-forces as the escape tower pulled her Mercury capsule away and did the equivalent of zero to twelve hundred miles per hour in under five seconds.

Until then she'd been doing fine on the lever-and-light test. The g-jolt blurred her vision for about thirty seconds and ruined her concentration. Then she got back to work while the capsule floated toward the Atlantic under its parachutes. When it smacked the water, she ignored the lever again for several seconds. Those two lapses were enough to set the medics off one more time. Instead of proving that an astronaut could function in a high-g emergency, it gave them ammunition to fret that he couldn't. That conclusion came later. Miss Sam came home to a hero's welcome and press acclaim. The monkey was every story's lead. All of the Mercury hardware worked perfectly too. But that didn't get mentioned much. The severity of the test, and Miss Sam's lapses, also ended any thoughts of letting Al Shepard or Gus Grissom ride a Little Joe. They'd have to wait for a more mature Mercury program.

Gilruth wanted that to happen in 1960, but soon knew that the schedule was too ambitious. The Mercury capsules they'd been using were "boilerplate"—close to the real thing, but not fully equipped. The first production capsule wouldn't be ready for a launch test until November. If everything went perfectly, maybe an astronaut could take a suborbital flight in January 1961.

Even that hope faded away after a major disagreement between Gilruth and the Air Force. July 29, 1960, dawned happily for NASA. President Eisenhower had approved a follow-on manned program, and the space agency announced that it would begin designing a three-man spacecraft that might even do a circumlunar mission in the mid-sixties. The program would be called Apollo.

But the Apollo news was buried in the crush of what came later in the day. It started sixty seconds after an Atlas missile, bought and paid for by NASA, left the launch pad at Cape Canaveral in July. A black Mercury capsule and an escape tower with its four angled red rockets rode on top. This was the first Mercury-Atlas flight, MA-1, the one that would "qualify" the capsule-and-rocket combination by proving that the two worked well together.

They didn't. Maybe it was the weather, maybe it was something else. The Cape was covered in clouds with stormy winds aloft. Walt Williams, trading the desert dryness for the humidity and turbulent weather of Florida, was overseeing his first mission as the Space Task Group's launch operations manager. He could have ordered a hold, or even scrubbed the mission until a better day. He didn't. MA-1 lifted off on schedule.

At T+60, beyond the clouds, out of sight of anyone on the ground, and without aerial cameras to record the details, the Atlas exploded in a ball of fire and smoke. Bob Gilruth was in the blockhouse slapping his head, and this time there was nobody immediately available to solve his problem. Pieces were still raining into the offshore surf when the Air Force decided that it was all Gilruth's fault. Or maybe Walt Williams'. Gilruth loosed a volley of criticism on Williams too, for being a slave to the schedule instead of factoring in the dirty weather. The January 1961 goal for putting an American into space was blown.

Williams took it hard. He had a metal plate engraved and it sat on his desk where he could see it every day.

> *You don't get medals*
> *for on-time failures.*
> *—Walt Williams*

Bob Gilruth had calmed down by the time the Air Force officially laid all the blame at NASA's feet. "The spacecraft came apart and went back and hit the Atlas and caused it to blow," Air Force investigators reported.

Gilruth looked them in the eye and held his temper. "Well, it might have," he answered, "but I don't think so." There were no data and no photography to support the Air Force position. Gilruth had another theory. "I believe the skin of your rocket is too thin," he said. The Atlas was little more than a stainless-steel balloon. Its skin was only one-fiftieth of an inch thick, and it was pressurized from the inside by its load of fuel. "The turbulence from our spacecraft is probably enough to cause it to oscillate and wrinkle," Gilruth said. "It probably ruptured."

No way, the Air Force said. The Atlas didn't blow up when it carried other payloads. Gilruth was adamant. "You can fix it by putting some belly bands around it," he said. "That'll stiffen it up."

That tore it. The Air Force was proud of its bird, and this civilian was insulting it by calling for *belly bands*. It sounded like something you strapped under the uniform of an overweight football player. "We paid for this rocket and we need to fly," Gilruth reminded the officers. The Air Force offered a compromise. They already planned to make future Atlas rockets out of heavier steel. It could take an extra five months. Maybe they could deliver one to NASA in July 1961. Gilruth and his team suspected that might be the rebuttal. "We can't wait five months," he said. "We need to put the belly band on."

The Air Force refused. Gilruth put all his cards on the table. "Okay, we'll take it to the Secretary of the Air Force," he said, and they met in Dudley Sharp's office in the inner ring of the Pentagon. Gilruth knew Sharp, had briefed him on other occasions, and trusted him. He presented his case, then tossed in the clincher. "I will agree to take all the blame if it breaks," he said.

Sharp understood the implications. Gilruth's job was on the line. He got his belly band. Mercury-Atlas 2 was ready in February 1961. There was a new Air Force secretary, Eugene M. Zuckert, and the day before the launch his press aide called NASA public affairs officer Paul Haney to dictate a "not our fault" news release. Haney had it in his pocket at the Cape Canaveral press site. If the Atlas exploded, he'd give it to reporters. The news release clearly put all the blame on NASA for insisting on this launch when it could have waited a few months for an Atlas with a thicker skin.

Gilruth was in the blockhouse a few miles away, and just before the countdown reached zero, he walked outside and stood behind a bush with his watch in his hand. The critical moment was T+60, when the Atlas reached Mach 1. That was when MA-1 blew.

It was chilly, but he was sweating. *Thirty seconds.* This grandfatherly middle-aged man, his head completely bald and the fringe over his ears as gray as Duluth snow, didn't care about himself. *Forty seconds.* He cared about Project Mercury. *Fifty seconds.* He cared about beating the Russians. *Fifty-five seconds.* He cared what people would think about his country if this one blew too.

Sixty seconds. I'll give it ten more. Seventy seconds. He wiped his brow. There was no need to slap it. *We're okay.* Bob Gilruth walked back into the blockhouse. "We can relax," he said out loud. "We won't have that problem again."

At the press site, Paul Haney ripped the Air Force news release into small pieces and dropped it into a trash can. He relaxed too.

That afternoon, some of the key people got rip-roaring drunk at the Holiday Inn in Cocoa Beach. With Gilruth on point, they'd taken on the Air Force and won. Charley Donlan was particularly excited and relieved. If Gilruth had lost, Donlan's job probably would go too. His boss wasn't there, so he drank enough for both of them. He finally passed out with a smile on his face. Walt Williams, toughened by his years in the desert with hard-flying, hard-drinking test pilots, and feeling good about the day's success, couldn't resist. He picked up a pitcher of martinis and poured half of it into Donlan's left ear.

Donlan twitched momentarily and kept on smiling. Williams shrugged and poured the rest of the booze into his own glass. NASA was getting tougher. It was starting to show some muscle.

The name Vostok was classified, a state secret known only to a few. The Chief Designer's team began sketching it out in 1958, and it looked nothing like the Mercury sketches being done simultaneously in Virginia. There was only one similarity between Vostok and Mercury. Each would carry one man into space.

That was Sergei Korolev's single-minded goal for Vostok. He didn't give a hoot about a cosmonaut doing anything while he was up there. If it was important to do work in space, that could come later. That included controlling the craft in orbit. Bob Gilruth gave his Mercury astronauts as much in-flight control as possible in the early days of space technology. Sergei Korolev gave his cosmonauts almost none at all. The two things a cosmonaut could control were among the secrets unknown to the West until decades later.

Mercury was bell-shaped. Vostok was a ball. Mercury had an escape tower powered by rockets to pull it free of a disabled or exploding rocket. Vostok had an ejection seat. If a cosmonaut had to eject off the launch pad, he'd be caught in a big net. Maybe. If he ejected any other time, he'd come down under a parachute. Both the Mercury and Vostok capsules had parachutes for recovery. But Mercury was designed to land in the ocean—a splashdown. Vostok came down on land—a crashdown.

Gilruth's Mercury designers listened to the medics and went with a pure oxygen atmosphere for the astronauts. That had several advantages and one big disadvantage. For breathing, the important factor is the "partial pressure" of oxygen. At sea level's atmospheric pressure of about 14.2 pounds per square inch of air, oxygen accounts for about 3 pounds. They made the pure oxygen pressure inside the capsule about 5 pounds per square inch, and that was generous. With less pressure inside the cabin, the Mercury capsule could be manufactured from lighter materials and it would be less difficult to seal against the vacuum of space. Every ounce lifted into orbit mattered; American rockets still lacked the brute power of Korolev's R-7 series. The drawback was fire. A fire in a pure oxygen atmosphere could be a disaster.

Korolev went with a normal atmosphere in Vostok. He could afford the extra weight. He didn't have to worry about the fire hazard. But he did need to include a pressurized space suit for his cosmonaut to wear during reentry —to fight off the possibility of nitrogen bubbles, *the bends,* forming in his bloodstream when the cabin depressurized at a certain point during reentry.

Tass announced the launch of *Sputnik 4* on a "scientific mission" on May 15, 1960. It wasn't a Sputnik. It was the first Vostok. No man or beast was on board and it operated well through four days in orbit. But when the Chief Designer ordered it down, things went wrong. Instead of setting up in the correct attitude so that its retrorocket would brake its motion and drop it from orbit, the Vostok aimed the wrong way. The retrorocket fired it into a higher orbit and there it stayed. Observers in the West noted the change in "Sputnik's" orbit and gave the Russians credit for extending the life of their craft. The retrorocket separated and finally came down in September 1962. The Vostok cabin fell to Earth in October 1965.

The problem was easily understood, so they tried again in July. This time Vostok carried two dogs, Chaika and Lisichka. The R-7M rocket exploded on the launch pad. Vostok was destroyed. The dogs died.

Exploding R-7s were nothing new. Korolev had a third Vostok, and more dogs, waiting. It took less than a month to roll a new R-7M to the pad with a Vostok on its nose. The launch was perfect. Tass called it Sputnik 5 and said that Belka and Strelka were on board. After a little more than a day in space, and data flowing that showed everything was working as planned, Korolev told his mission control team to signal it down. Two hours later the Vostok dropped to Earth under its parachutes and the dogs came home. They were the first living things to go into space and come back alive.

Among the watchers when the Sputnik/Vostok lifted off were a half dozen relieved young men. They'd been there when the R-7M exploded in July, killing Chaika and Lisichka, and each of them imagined himself in that fatal inferno. Now with Belka and Strelka safely home, they breathed another sigh of relief and turned back to their duties. The six were the first real cosmonaut trainees. They were small men, 5'6" or less, and none weighed more than 143 pounds. Vostok's interior was cramped; the ejection seat and its mechanisms took up more room than a cosmonaut. When the call went out for volunteers, it specified that only small men need apply.

These six—Valery Bykovsky, Gherman Titov, Grigory Nelyubov, Andrian Nikolayev, Yuri Gagarin, and Pavel Popovich—survived the medical tests and were chosen. They were jet pilots, but not test pilots, and each was a superb physical specimen. They also were smart. The Chief Designer had little need for their intelligence in surmounting the initial challenges of putting a man into space. His own people dealt with those problems. For the early flights, Sergei Korolev wanted no more than tough men who would survive. Later on, maybe, having experienced cosmonauts with good brains might be useful. But not now.

The cosmonauts traveled in the spring of 1960 to a rail junction in the Siberian steppes where a green bus and its driver waited. A few hours later they arrived at their new home, the Tyuratam base, where the Chief Designer mounted his attack on outer space. In Soviet announcements, the remote space center was called Baikonur, probably in a futile attempt to disguise its true location. It was closer to the city of Tyuratam and the ruse didn't work; the names were used interchangeably in the West, and anyone who mattered knew exactly where the launch pads and training facilities were sited.

Training was gut-wrenching, more physical than intellectual in the months that followed. They flew airplanes and learned gymnastics, lifted weights, ran until their hearts and lungs were conditioned beyond anything they thought possible. Some days were devoted to classroom studies. Physicists explained the realities of space to them—the airless vacuum, cosmic radiation, zero gravity. Doctors who had studied the data from the early dog flights told them what they could about how living things reacted up there. Psychologists prepared them for the isolation they'd feel and watched their reactions in class and then in sensory deprivation chambers where they sat alone for hours at a time in silent darkness.

The Chief Designer himself introduced them to a Vostok capsule. They sat inside the big round ball, looked out through its thick glass portholes, and imagined themselves seeing a blue-green world passing below. Korolev explained that the chosen man wouldn't have much to do during the first single-orbit spaceflight. He would talk on the radio, of course, reporting instrument readings and his own feelings and describing what he saw inside Vostok and outside. He could change settings on the air-conditioning system if he was too hot or too cool, and there were backup switches and systems if a primary piece of equipment failed. He pointed to a control panel with six numbered buttons. That was the emergency retrofire system. If the automated system failed and Vostok's retrorockets didn't fire, the cosmonaut could enter a secret code that would make them work.

Here was the place where Soviet paranoia and distrust of individuals reared its ugly head. The cosmonaut would know three of the numbers, Korolev explained. If the worst happened, someone would radio the other three numbers to him. The hidden agenda was unspoken but obvious. There was no way for the cosmonaut to defect from space. He couldn't just decide to come down in America or Europe and make it happen by entering the six numbers. And if he did try, there was a fail-safe even beyond the numbered buttons, a cache of explosives that could destroy the spacecraft on a signal from the ground.

None of the cosmonauts asked what they should do if the radio wasn't working. Some questions were best unanswered.

The retrofire scheme was an obvious difference between the American Mercury program and the Vostok program. There was another too, even more dramatic. Each of the cosmonauts was put through a new round of parachute training. They'd all jumped a few times as part of their pilot schooling. That was mandatory in the Soviet Air Force. Now they jumped, and jumped often, because getting out of a Vostok crippled high in the atmosphere might be required. For that contingency, they learned the techniques of free fall and of delaying their parachute opening until they reached the denser air below.

None of them liked the idea of ejecting from Vostok right off the launch pad and being caught in a net. But they learned to enjoy leaping from high-flying airplanes, spreading arms and legs to slow themselves as they fell, and finally pulling the cord that gave them a billowing parachute over their heads and a safe landing below. They enjoyed parachute days, especially when the option was running for miles across the steppes or sitting alone in a deprivation chamber until all sense of time and space faded away.

And always they were watched. One of the six, Sergei Korolev promised, would be chosen. There was still work to do. But the Chief Designer had his eye on the spring of 1961. He felt good about it after reading the American plan for Mercury: A lowly suborbital flight was to be first with a man. Man-rating the Atlas rocket was a longer process than anyone thought. And other tests needed to be done. Before an astronaut went in orbit, one of them would be launched in a Mercury capsule by one of Wernher von Braun's Redstone rockets. The Redstone could put Mercury into space, but it didn't have the power to keep it there. It could provide just enough velocity for an up-and-down flight of maybe fifteen minutes. Still, the astronaut aboard would have gone into space.

Sergei Korolev was ready to rebut that claim. He had esteemed Soviet scientists ready to make statements, and technical papers already written and available to be passed out to the press. If an American flew a suborbital flight before Vostok was ready, the Chief Designer would make certain that the world understood. This was a gallant achievement, Korolev's scientists would say, and the Americans are to be congratulated. But it was not a spaceflight. Not at all.

The eighth astronaut wasn't happy. But when Al Shepard heard him being interviewed, he told the other six and they all cracked up.

The Original Seven reluctantly put up with interviews everywhere they went. Number eight was not reluctant at all.

The seven got a bit of protection for themselves and their families in late 1959 by signing a contract with *Life* magazine that paid them $500,000 a year, split seven ways, for their personal stories. The rest of the press screamed bloody murder at the *Life* exclusive, but NASA ruled that the contract did not cover official duties. It gave *Life* exclusive access to the astronauts' families and home life, to their private activities, and in a real stretch, to their impressions and personal observations. *Life* also provided life insurance for the seven, who couldn't get policies any other way. The justification that *Life* was helping to provide for astronaut families if a space tragedy occurred helped to get the deal approved.

In practical terms, astronauts were required to give full and detailed reports to NASA after space missions, and as test pilots they did this enthusiastically. Then they would have a lengthy and less enthusiastic press conference to answer reporters' questions. Finally they would sit down in private with *Life* writers, who would ghost first-person accounts of the mission for the astronauts' bylines. These sessions often were the least enthusiastic of all. Ralph Morse was usually there to record the event. He and other photographers filled *Life*'s pages with pictures of the wives and kids, and of the astronauts with their cars, sailboats, on vacation, and whatever else they could find to show how these men lived.

Morse was an ebullient fellow with a New York/New Jersey accent, and the astronauts genuinely liked him. Some of *Life*'s writers, who were mostly around for the mission highlights when tensions were high, were only tolerated. But the astronauts all understood two important facts. They were being paid handsomely for *Life*'s intrusion. And the contract meant that their families were spared the onslaught by the rest of the media. They could decline interviews and photo sessions without criticism, which was leveled at *Life* instead.

The eighth astronaut first told his story on *The Garry Moore Show* in early 1960. His name was Bill Dana. That wasn't the name that became a fixture in the space race and beyond. His interview on *Garry Moore* had the nation rolling on their living room floors.

Interviewer: [We have with us] the chief astronaut for the United States Interplanetary Expeditionary Force. How do you do, sir. May we have your name?
Astronaut: My name . . . Jose Jimenez.
Interviewer: You're the chief astronaut with the United States Interplanetary Expeditionary Force?
Jose: I am the chief astronaut . . . with the Hew-nited States . . . Interplanetar . . . My name Jose Jimenez!
Interviewer: Mr. Jimenez, would you tell us a little about your space suit?
Jose: Yeah. It's very uncomfortable.

It was all done in a hesitating mock-Hispanic accent. This astronaut was obviously not up to the standards of the Original Seven.

Interviewer: And what is this called? A crash helmet?
Jose: Oh . . . I hope not!
Interviewer (attempting to recover): Mr. Jimenez, you must have some opinions on the race for space.
Jose: All right, I will . . . I got one.

But the silence was deafening, filled only with a nation laughing uncontrollably. Whatever Jose's opinion, he kept it to himself. The routine went on, finally getting to the real point of it all.

Interviewer: Mr. Jimenez, do you have any message for the American people?
Jose: Yeah, I got one . . . Please! . . . Don't let them do this to me!

Jose the Reluctant Astronaut was a smash hit. Bill Dana's career took off . . . like a rocket. And Al Shepard couldn't stop laughing. It took a few months, but the eighth astronaut and the Original Seven would form a friendship that added an intensely human and very funny touch to the space race that none of the participants could have predicted.

Dana took his act on the road, signed to appear at the "hungry i" nightclub in San Francisco, and got a record deal from producer Mickey Kapp of Kapp Records. He had other characters, and other Jose shticks, but it was the Reluctant Astronaut that crowds wanted to see. Kapp set up his equipment at "i" to do a live album, and the astronaut bit knocked them dead. Kapp turned it into a single record, had the album's cover art redone to feature Jose being launched into space, and had a hit.

He sent eight copies to Lieutenant Colonel Shorty Powers at the Space Task Group headquarters in Langley, Virginia, with a note asking Shorty to keep one for himself and give the rest to the Original Seven. He never got a response.

Then Bill Dana took his act to a nightclub in Cocoa Beach, Florida. The space folks there ought to be a good audience. His phone call to Mickey Kapp after his first set was all but hysterical.

"Mickey, you gotta get down here!" Dana screamed to Kapp in New York

City. "You wouldn't believe it! I'm just starting the Jose bit when this guy jumps up on the stage and took over as the interviewer.

"He knew the whole thing, Mickey! The place went wild. Mickey, it was Al Shepard! And he had Deke Slayton and Wally Schirra there too. They knew every word!"

If the astronauts had *Life* to protect them from the press, other enemies could make life miserable for Bob Gilruth, and he paid scant attention to what the Russians were doing. It was the Americans that worried him.

Neither of Dwight Eisenhower's science advisers, James Killian, then his successor, George Kistiakowsky, had any fondness for putting men into space. Both thought Mercury was a gimmick and that its budget could be better spent on other kinds of science. Gilruth's position as director of the Space Task Group put him in frequent contact with the White House and with other major science groups. "Some of those people were just awful," he said. He was deeply offended in one meeting when Kistiakowsky put down the Mercury program in brutal terms.

"It will be the most expensive funeral a man has ever had," Kistiakowsky said, and Gilruth bristled with anger. Astronaut safety was close to a religion with the STG team. Gilruth and his people understood the dangers faced by test pilots, and soon to be faced by astronauts, and spent whatever it took to mitigate them. When Kistiakowsky kissed off their efforts with a prediction of death, Gilruth cut him out of the loop.

It wasn't hard to do. The White House science advisers were pissing into the wind of politics and public opinion. Eisenhower had been roundly criticized in the press for his slow response to the Russian space threat and for not leading the charge into space. That *Newsweek* story about probably fictitious cosmonauts in October 1959 was preceded a week earlier by another element of historian Walter A. McDougall's "media riot"—a story headlined "How to Lose the Space Race!" *Newsweek*'s Edwin Diamond wrote that "no amount of soft soap can gloss over the dismal fact: The U.S. is losing the race into space, and thus its predominance in the world." That kind of word riot was exactly what the Chief Designer wanted to see in American newspapers and magazines.

Two of the items on *Newsweek*'s list of Eisenhower faults were "start late" and "think small." He had been guilty of both. But no matter what he thought in private, he was not about to make those mistakes again by shrinking or canceling Project Mercury. He'd even given his approval—grudgingly in private, but enthusiastically in public—to the follow-on Apollo program.

With his ears still burning from Kistiakowsky's insult, Gilruth tromped over to NASA headquarters on Lafayette Square to see the agency's deputy administrator, Dr. Hugh Dryden. "I'll never go back to Kistiakowsky again," Gilruth said after telling the story.

"You don't have to," Dryden said. "He doesn't have the say anyway. Don't worry about Kistiakowsky. You did fine. I think you're going to have no trouble with that group of scientists."

He didn't. But the 1960 presidential campaign and John F. Kennedy's narrow victory gave NASA a new set of troublemaking scientists. The space race and the missile gap were issues, and Kennedy didn't hesitate to blast Eisenhower and Nixon during the campaign. He'd been advised during the campaign by a science committee chaired by Jerome B. Wiesner of the Massachusetts Institute of Technology. Wiesner was another of the anti-scientists on the issue of manned spaceflight, and he had Kennedy's ear. He would complain plenty after the election.

The genius of Bob Gilruth and his manned spaceflight team only became obvious later. He understood that the advance work took time and must be done right. He set superhuman goals, and his dedication flowed downward to the riveters and wiring harness makers. Then when it was time, the action began and it was virtually nonstop. Even Gilruth didn't know it in the fall of 1960, but an aerospace team unlike any team ever fielded anywhere in the world was forming up in the United States. Its hallmarks were brilliance, innovation, passion, and sacrifice. The only clock that counted was the countdown clock.

The McDonnell Company that fall had thirteen thousand people assigned to Mercury, with the capsule production and assembly crews working three shifts seven days a week. Gilruth's own Space Task Group people, government employees, had virtually stopped taking vacations and were working fifty- and sixty-hour weeks. Launch crews at Cape Canaveral worked seventy-hour weeks. The biggest complaint was that there were not enough hours in a week.

It was no different at General Dynamics in San Diego, where the Atlas was built; at Rocketdyne in Los Angeles, where rocket engines flowed down assembly lines; at the new Marshall Space Flight Center in Huntsville, Alabama, where Wernher von Braun's people and their industry contractors built Redstone rockets; or anywhere else connected with Mercury. This was a race, the only race, and there was unparalleled dedication among the people running it. A day was coming when Bob Gilruth could stand at his office window, close his eyes, and see its runners numbered in the millions. Their passion overwhelmed every obstacle.

When politicians and the press said that Mercury was lagging, Gilruth's answer was one word: "Nonsense." His team planned to launch two unmanned Mercury capsules with Redstone rockets before putting an astronaut on the third suborbital flight. If everything went perfectly, the MR-3 astronaut flight could happen in early 1961. The first Mercury-Redstone, delayed by problems during technical checkout and by weather, went up eleven days after the election. The problem was, it didn't go very far up. Jerry Wiesner could cackle and point to a four-inch flight as the best Mercury could do. He wasn't the only one laughing. It was a bitter and funny end to an important mission.

The countdown for MR-1 was perfect. Its Rocketdyne engine belched smoke and fire. The rocket itself shook and shimmied, lifted about four inches off the pad, and as the engine quit, settled back down. What happened next left everyone shaking their heads. Mercury's escape tower fired off but didn't take the capsule along. Alone and empty, it soared to four thousand feet before its little rockets burned out. Then it fell back and crashed a few hundred yards from the pad.

While that had people's attention, they heard another loud *pop*. With no escape tower clamped on and a scramble of electric signals flowing through its wires, the capsule shot its small drogue parachute into the air. The little chute opened, and as it floated down, it did its job perfectly: It pulled the big main parachute out of Mercury's upper section. The reserve chute followed in a few moments. America's hope for being first to put a man into space stood on the launch pad, draped in its parachutes like an uncertain child hiding under a blanket. It was not Mercury's finest day.

At the same time, Mercury's recovery beacons turned and began transmitting. Far offshore, rescue aircraft picked up the homing signals and turned toward shore. Dye-maker canisters aboard the Mercury capsule popped and fell to the pad's concrete base. The next *pop* was from smoke bombs, spewing their attention-getting signal. It was a comedy of serious proportions.

In the launch center viewing room, Paul Haney of the public affairs office turned to Max Faget. "Give me a sequence, Max," he said. "Is that the way things are supposed to happen?" Faget fixed him with the evil eye, so Haney explained that they needed a story for the press.

"Why do you have to tell them anything?" Faget snapped. "We didn't have a flight!"

But they did have to say something. Reporters had seen it all and at that moment were staring at a forlorn rocket that hadn't done its job. There'd be a press conference in an hour. And if Kurt Debus had his way, they would have had a bigger story than anyone thought. Von Braun's resident Redstone expert at the Cape, with the title of launch director, had a wild scheme in mind. Walt Williams took the call at his operations director's console in mission control.

"I've called for a man with a gun," Debus said in his thick German accent.

Williams wasn't sure that he'd heard it right. "A gun?"

"Ja," Debus confirmed, and said he wanted a rifleman to shoot holes in the Redstone's liquid oxygen tanks, allowing the stuff to drain out harmlessly before he sent crews to the pad to render the rocket safe.

With Bob Gilruth now at his side, Williams put a permanent hold on that idea. The thought of reporters watching a NASA sniper taking potshots at the rocket was more than either of them could handle. They could see the Redstone parachutes rippling in the breeze and decided to let everything just sit for a while. As the liquid oxygen (LOX) warmed up, it would push open vent valves and escape anyway.

The press conference was tense and the worst possible question was one of the first asked: "What if there was an astronaut in the capsule? Would you still be sitting here waiting for the LOX to evaporate?"

"No," Williams answered, shivering inside at the danger involved, "we'd get him out with the cherry picker."

They traced the problem to an electrical plug. It was supposed to pull out as the Redstone began to lift up. One prong was slightly longer than the other, and that was enough to do the dirty deed. On a Redstone military missile, the time difference between the two prongs cutting off the flow of electricity didn't matter. But on a Mercury-Redstone, considerably heavier, the plug came out more slowly—it took twenty-one milliseconds to disengage. That was enough for the computer to think that the main rocket should be shut down. So it did. On such minutiae rested America's hopes in space.

They tried it again a few days before Christmas and MR-1A was perfect. But they still needed two successes, and now both MR-2 and the astronaut flight on MR-3 had slipped into 1961.

Kennedy had won the White House in one of the country's tightest elections. Nine days before he was sworn into office, JFK announced that Jerry Wiesner would take the job of White House science adviser. The news cast a pall over NASA. The administrator, T. Keith Glennan, was on his way out. No one had a clue as to whom Kennedy would appoint, but if Wiesner was his science adviser, the future didn't look good. Then it got worse. JFK's transition team released a report prepared by Wiesner and a group of other scientists that blasted both NASA and the Pentagon. It was naturally critical of almost everything that Eisenhower had done, or not done, regarding missiles and space. Bob Gilruth blanched when he read the sentence "We should stop advertising Mercury as our major objective in space activities."

The next day's newspapers made it seem even worse. Reporters assumed that the Wiesner Report was Kennedy's policy. It recommended all kinds of changes at NASA and in the Mercury program—everything from scrapping the Atlas as a man-rated booster to rethinking the whole idea of putting a man in space. Killing an astronaut, Wiesner said, would be a national tragedy. It would. But abandoning space to Soviet cosmonauts would be worse. That was a fact of domestic political life and of international diplomacy that Kennedy understood, even if Jerry Wiesner didn't. Releasing the Wiesner Report without background briefings on what JFK's policy would be, or at least with some caveat that would give reporters pause, was a mistake.

President Kennedy would continue to support Mercury. When that became obvious almost immediately, NASA shrugged off Jerry Wiesner and went back to work.

Reporters had a lot to learn about the intricacies of space. Covering a story as complex as a space mission was not like covering a political campaign or even a scientific meeting. The details, some of them obscurely embedded in engineering, made the difference between a mission success and a mission failure.

At the beginning of the space race, few reporters had the background or knowledge to write accurately detailed stories. Many of them would learn, and one of the positive fallouts of the race was that a corps of smart and skilled journalists would develop and become aerospace experts.

Then there were those who didn't care.

Lieutenant George Alexander, then a public information officer with the Air Force Ballistic Missile Division, was in the blockhouse during an Atlas launch. He watched the incoming data all abruptly go to zero and stop. The rocket had exploded downrange, out of sight of anyone at Cocoa Beach.

The expletives were still echoing in the room when Alexander went to the phone to make the calls to the press announcing another failure.

Then he drove into town, stopping off to see Howard Benedict of the Associated Press in his office at the Cape Colony Inn. Benedict was dealing with a query that disputed his launch failure story.

A local stringer for some out-of-town newspapers had filed a story to the *New York Daily News* saying that the air force announcement was a hoax. The Atlas had not exploded; it had sent a secret satellite on the way to the moon, Doug Dederer reported.

Alexander was dumbfounded. He'd never heard of Dederer until that moment. "Where is this guy Dederer?" Alexander asked the AP reporter.

"Go to Ramon's," Benedict said. "He'll be the guy in shorts and a madras shirt, smoking a twisted cheroot and drinking Jack Daniel's, sitting at the end of the bar with one knee up."

Ramon's was only a half mile away. Alexander charged over and found a man who fit Benedict's description exactly.

"Are you Doug Dederer?" he asked the seedy guy at the end of the bar. The man slowly turned and looked at the young officer with disdain.

"Yeah," he said at last. "And who are you?"

"I'm Lieutenant Alexander from Air Force public affairs," he said. "Howard Benedict says you've filed to the *Daily News* that our Atlas didn't blow up tonight . . ."

Dederer cut him short. "What's your problem?"

"I don't have a problem. You do. It blew up. It's not headed to the moon. I was there. Your story isn't true."

Dederer thought about it for a moment, but didn't look up again from his drink.

"My story'll read better than yours," he said. "Beat it, kid."

There were legends made in coming years at Cape Canaveral and Cocoa Beach. Doug Dederer's reporting was not among the savory ones.

Six men in Siberia and seven men in Virginia sat down to do the same job. Bob Gilruth had his seven do it in December 1960. Sergei Korolev waited until April 1961 to tell his boys what he wanted them to do. Both Gilruth and Korolev wanted a peer review. The day after MR-1A flew a perfect suborbital mission, Gilruth called his astronauts together and told them to start thinking about who was going to be first.

They'd been thinking about it since the day of that first press conference. For twenty-one months, they'd been training, competing, and becoming friends. Each of them knew, beyond any doubt, that he was the best of the seven. The friendship thing was a little troublesome. They liked each other and now they'd done enough carousing and skirt-chasing together to have a solid idea of what the next guy was made of. Maybe John Glenn wasn't quite in the inner circle with the other six. He could be preachy and sanctimonious, and when their drinking and womanizing threatened to become public knowledge, he'd lecture them about God, country, and image. He'd already been tagged in the press as the "old man," the astronaut father figure who understood the nation's need for heroes.

But he was good. It was the preachy part that rubbed them wrong. If they could stay square with their wives at home, what business was it of this Boy Scout marine when they were ready to party? It was almost a reasonable question. Years later, when Louise Shepard faced the question of Al's dalliances from unthinking friends or cruel enemies, she just shrugged and smiled her standard answer: "What do you expect from a sailor?" The flip side was that America expected more from these sailors and fliers. They got it straight up from the marine, and when the other six didn't walk the line, the press covered for them and didn't tell nearly all that it knew.

So when Bob Gilruth told them to think about who should be first, six of the seven didn't think it should be John Glenn. And when he told them to go a step further—*except for yourself, write down the name of which other guy should be first*—it's almost certain that Glenn's name was not on any of the slips.

Deke Slayton was glad to be there at all. He knew that he was the best of them all, but a few months earlier he'd gotten a scare. While the medics were wiring him up for a centrifuge run, they saw something. The monitor showed Slayton's heart had an irregular beat. It hadn't been there before.

They replaced the sensors and wiring harness and it went away. Slayton assumed it was an equipment malfunction.

But when they started the centrifuge, there it was again. *Thump-thump-athumpthump-thump-thump.* They stopped the run and for the next month looked seriously into Slayton's arrhythmia. He was grounded until this thing was checked. It was rarely there. Slayton threw himself into physical training, putting every stress he could on his heart to make the *athumpthump* disappear forever. His blood pressure was fine. He showed no sign of heart disease, arterial clogging, or any other cardiovascular dysfunction. There was just that random *athumpthump* and it showed up often enough that the doctors heard it.

Slayton understood the implications. This damned *athumpthump* wasn't going to kill him, but it might as well. Because if the doctors ruled against him, he would no longer be an astronaut. He would no longer be a test pilot. He would no longer be any kind of pilot. The fastest and most dangerous thing he could operate was a car. He'd rather be dead. This time he was lucky. He was more fit than any athlete, with quicker reflexes, more strength, and greater endurance. Whatever was causing the heart murmur was not interfering with his duties. The doctors cleared him to fly and to still be an astronaut.

A month after the peer review, Gilruth asked his boys to stick around for a few minutes after work. A new rocket and capsule, Mercury-Redstone 2, was being readied at Cape Canaveral. This one would have a chimpanzee on board. A good ride meant that an astronaut was next in line. They waited in their office, seven men sitting at seven steel desks crammed into a single room. At a little after 5 p.m. on Thursday, January 19—Kennedy would be inaugurated the next day—Bob Gilruth walked in looking uncomfortable. He passed his hand over his bare pate, stood for a moment to let the casual conversations among the seven go silent, then told them that he'd made a difficult decision.

"Shepard gets MR-3 and Grissom gets MR-4," he said in a tenor voice that showed strain. "Glenn backs up on both flights."

This was a closely held secret, he warned them. There would be no public announcement "until the right time." If the office were a locker room and the seven were a ball club, there would have been whooping and hollering and shouting and backslapping. Someone would have popped a champagne cork and after dousing Shepard and Grissom and Glenn, would have turned the bottle on Bob Gilruth.

Nothing like that happened. There was silence. Four of them felt like they'd been slapped. *I'm not even in the top three!* Two of them experienced momentary glee before the other thought surfaced hard and fast. *I'm not going to be first . . .*

Al Shepard's mind went blank. Then his blue eyes twinkled and the corners of his mouth turned up in a slow grin. *I got it . . . I got the first ride!*

It wasn't silent for long. Bob Gilruth held out his hand and Shepard shook it. The others woke up and joined in. These seven were hard men and dedicated. They'd experienced terror, known disappointment, visited with death. A little thing like this wasn't the end of the world. And there'd be orbital flights as soon as Shepard and Grissom had their rides. Now that was an assignment worth competing for!

From liftoff to splashdown, Al Shepard's ride was flight-planned to last only fifteen or sixteen minutes.

Someone had decided that the atmosphere ended and space began at 300,000 feet. That's the official border. Cross it and you're in space. Shepard would hit 300,000 feet in less than four minutes, continue up to 615,000 feet, about 116 miles, then arc down through reentry and landing. But he couldn't go until Ham did it first. "First the chimp, then the chump," a wag put it. Bob Gilruth laughed at the joke, but Shepard didn't think it was funny.

Ham, named for Holloman Aerospace Medical Center in New Mexico where he'd trained, was the doctor's last-ditch try at proving or disproving those onerous theories about man's capacities in space. They'd devised a diabolical device to test him during his ride.

He had electrodes pressing the soles of his feet. Across his lap, a small control panel let him stave off shocks. He had two lights and two levers. When the white light came on, he had fifteen seconds to push the right-hand lever. If he didn't, he got shocked. When the blue light came on, he had just five seconds to push the left-hand lever. The blue light only flashed every two minutes. If he reacted right, he not only avoided the shock but got a banana pellet. The test ran from liftoff to splashdown. So Ham had to perform during the high g-loads of launch, the weightlessness of space, and the even higher g-loads of reentry. If Ham did his job under all those conditions, the medics reasoned, so could Al Shepard. None of the astronauts liked the comparison. They had no doubts at all about their ability to perform. And the Russians were breathing hard to get their man into space first. But they had troubles of their own. Shepard could smell victory for the United States, if only Ham did his job.

It might have worked out. On the other side of the world, Sergei Korolev was having a bad run. Rocket production was increasing, and because he now had priority authorization to get a man into space—Khrushchev fully understood the international impact this would have—the military had to stand aside while the Chief Designer took whatever number of R-7s he needed. So when he got another dog-carrying Vostok into orbit in early December 1960, then saw it burn to cinders in the atmosphere when its control system failed, he quickly ordered another rocket to be readied.

That one was wheeled into place, pushed upright from its carrier, and launched on December 22. This time the R-7M's third stage failed. Vostok's lucky two-dog load survived because it was high enough in the atmosphere to spring free of the faltering rocket, open its parachutes, and proceed to a rough but nonfatal crashdown.

The Vostok program was not inspiring great feelings of confidence. Korolev had now launched five of them, with only one success. Two failures were blamed on the launch vehicle. So that left three. One of those misfired at reentry time and went into a higher orbit. One worked perfectly. And the third misfired, came home all wrong, and burned up. The control system that pointed Vostok for retrofire was the culprit. Korolev ordered a complete stand-down until his engineers and technicians pinpointed the problems and fixed them.

It looked like the Americans could finally win a big one, even if it amounted only to sending an astronaut up and down on a suborbital flight. The Chief Designer didn't count on the unwitting help he got from a past and present enemy.

They strapped Ham into his custom couch, trundled him out to the Redstone launch pad, made sure that his personal test equipment, including his banana pellet mechanism, was working, and sealed him up in the Mercury capsule just before 8 a.m. It was January 31, 1961, just another day's work for the well-trained chimp. The countdown was normal—it had one problem after another. Walt Williams was in the control center; he'd moved back from the desert to become operations director for Mercury. But the man in charge was Chris Kraft.

With each mission, success or not, they were evolving their procedures, and one rule became first obvious, then inviolate: Final authority rested with the launch director until liftoff, and then with the flight director. In the stress and strain of an ongoing mission, the flight director could not be overruled by anyone up to and including the president of the United States. He could be fired later, but he couldn't be challenged in the heat of a crisis. Walt Williams made Kraft his flight director, and he would hold that job through Mercury, and then much longer. On the control center and tracking network communications loop, each person was named for his job.

The astronaut who was the sole voice link between the ground and the capsule was Capcom. The flight dynamics officer, who monitored trajectories and in-flight propulsion, was Fido. The medical officer was Surgeon. The specialist who monitored the onboard electrical, environmental, and communications systems was EECOM. The expert charged with calculating everything down to a fine point for retrofire was Retro. The person on the loop at each of the worldwide tracking sites, except for astronauts deployed to some sites during Mercury, was known by his location. He was Woomera or Hawaii or Canary or CSQ, for the ocean vessel Coastal Sentry Quebec. There were others, and as missions became more complicated and expanded beyond Mercury, their numbers grew.

The flight director was then and forever Flight. He had control and he was the focus of all talk on the mission director's voice loop. The Capcom could talk to Flight, or switch to the capsule channel and talk to the astronaut. Capcom conversations also were heard by all. Everybody else talked only to Flight on the director's loop. On separate voice loops, each console could talk to supporting staff in back rooms or other locations. But the primary loop was for Flight. He queried, he listened, he made the final decisions, he issued orders. In tight situations, he'd do the whole process in a matter of seconds. Being Flight was never for the faint of heart.

Christopher Columbus Kraft, Jr., was Flight in Mercury and became a true American legend. From then on, he would be "the man called Flight," and when you say it that way, certain segments of NASA, from the youngest at their consoles in the twenty-first century to men long retired with their pensions and their memories, know who you mean.

Launch time was supposed to be 9:30 a.m. When the troubles started, the clock stopped. One of the problems was an electrical box that was overheating. Kraft ordered a hold, then worried about Ham's little space suit. If the chimp overheated as much as the electrical component, he'd have to decide whether or not to scrub the mission.

EECOM told him that Ham's suit temperature was holding in the comfortable range. Surgeon responded to Kraft's sharp questions with a quick evaluation of Ham's ability to withstand heat. Kraft nodded and made his decision. If the component cooled off some, the mission was go. It did, but then overheated again. Kraft asked for more reports on Ham. The chimp was hanging in there. They waited another hour for the component to cool. Kraft polled the room. It was getting late, almost noon before they could count down to zero.

Ham still looked good. This wasn't a mission where Flight could ask for a report straight from the astronaut. "Pick up the count," he ordered. Then he stopped it again when an elevator at the pad stuck and when other little problems cropped up. One piece of Flight's mind stayed focused on the capsule and its passenger. He was exhibiting a trait that was picked up by everyone who worked for him and that every Flight who followed accepted without question. The trait was, *The crew comes first.* It didn't matter to Kraft that this crew was a chimp. He didn't know how to differentiate. Ham was *crew*. A control center reflex took form that day and was refined to a purer form on every mission that followed.

MR-2 lifted off at 11:55 a.m., and what happened next was out of Flight's hands. But even if the crew was human instead of chimp, he wouldn't have changed anything. The plan called for the Redstone to take the capsule to 4,400 mph and 115 miles high before the escape tower pulled the capsule away and it descended to splashdown. Ham would endure a pressure of about 9 g's during the boost phase, then would go weightless for almost five minutes. During reentry he'd feel the force of almost 12 g's pressing him into his couch, but only for a brief time.

Through it all, Ham was supposed to watch those lights and toggle those levers. If he didn't, he'd get zapped through the soles of his feet. And if he did, he'd get a banana pellet. If Ham could do all that under such extreme conditions, the medics were fairly certain that Al Shepard could too, and without the bother of zapping his feet or giving him banana pellets.

It looked good to reporters and to the thousands of winter tourists on the beach as MR-2 roared and crackled skyward. In mission control, Flight was getting a different message. Within a minute and only a few miles out over the Atlantic, the Redstone rocket was slightly off course. It was climbing just a little too steeply and the angle was getting worse. It wasn't enough for the human eye to see, but the computers knew and transmitted the data to mission control.

"Flight, FIDO."

"Go FIDO."

"One degree high . . . projecting a 17 g peak load."

"Are you go?"

"Go, Flight."

That was all Chris Kraft had to hear. He did nothing.

"Flight, FIDO."

"Go FIDO."

"LOX depletion in twenty seconds."

Kraft looked at the clock. The steep climb was using up the liquid oxygen, LOX, portion of the Redstone's fuel faster than normal. He had only seconds to make a decision: abort or don't abort. When the LOX tank ran dry, he knew that the automatic abort system would take the decision out of his hands anyway.

"Go or no-go, FIDO?" demanded the man called Flight.

"Go, Flight."

"How's the crew, Surgeon?" Flight asked.

"Responding well, Flight. Crew is go."

Kraft stood by his console with his arms crossed and waited.

"Flight, FIDO, LOX depletion . . . engine shutdown . . . abort initiated . . . we're go, Flight."

Kraft nodded. There was nothing to say and nothing to do but watch the computer plots and the console instruments.

"Flight, EECOM."

"Go, EECOM."

"Cabin pressure's falling, down to one psi."

"How's the suit circuit?"

Ham's couch was sealed in a pressurized unit. In manned flight, the suit circuit maintained a flow of pressurized oxygen to the astronaut's space suit. In the monkey flights, it kept the couch unit supplied with oxygen.

"Holding at 5.5, Flight."

So somehow a valve had opened and let most of the pressurized oxygen out of the capsule. But the suit circuit was doing its job. If Al Shepard had been aboard, he would have been sealed in his space suit during launch and reentry. He would have known there was a problem, but he would have been safe and secure. Ham was safe and secure too, didn't know there was a problem, and just kept working the levers.

"Flight, Recovery, we've got a beacon."

"Rog, Recovery."

The capsule's recovery radio transmitter had turned on when the escape tower fired to pull it away from the now-dead Redstone.

"Flight, Recovery, we're going to overshoot."

"How far?"

"Maybe four hundred miles, Flight. I'll know more in a few minutes."

"Tell the ships."

"We just did, Flight."

"Good job, Recovery."

The tracking data and telemetry from the Mercury capsule was flooding in to mission control. The steeper climb was sending Ham up to 157 miles. By the time they had this calculated, he was over the top and coming back down. He'd been slammed with 17 g's of pressure during the ascent and he was getting nearly seven minutes of weightlessness instead of the five minutes they'd planned. Now he was enduring a 14.7 g-load during reentry and would splash down 132 miles beyond the target point.

Nobody'd told Ham about the plan, so he didn't know the difference. He just kept doing his job.

Mercury's chutes opened perfectly and the capsule hit the water seventeen minutes after liftoff. But Ham's ordeal was just beginning. It took a Navy plane a half hour to find him bobbing in the Atlantic. Helicopters arrived nearly an hour later and the sight was frightening. The capsule had tilted onto its side and was sinking. Somehow Mercury was taking on water. Jumpers went into the sea, attached cables, and as a pilot applied full power to his chopper, they lifted it slowly into the air. They got it back to the *Donner,* lowered it to the deck, and freed Ham. When engineers examined the capsule, they found holes in the bottom where the heat shield smacked it too hard on splashdown. As it took on water, it tilted over until the air valve that had opened when it was still on its way up let water in from that end too. Ham was dry in his couch unit, but he came close to ending his brief space career on the bottom of the Atlantic.

There was a movie camera on board, plus all the readings from Ham's workbench. He was near-perfect in pulling those levers. His left-hand performance was 100 percent. He was slow a couple of times with his right hand and got a few electrical shocks. But overall, through the extra g-forces of launch and landing, and the 6.6 minutes of weightlessness, he left them applauding. He wasn't even upset at the recovery delays and near-sinking. He just grinned and gobbled down an apple and part of an orange when they set him free. The medics made it official. They weren't totally convinced and a few would make trouble later, but the word was out. If Ham could do it, so could Al Shepard.

The problem now was Wernher von Braun. He sealed Al Shepard's place in history and made the Chief Designer a happy man.

President John F. Kennedy appointed a dyed-in-the-wool southern bureaucrat to be NASA's new administrator. He was fifty-three-year-old James E. Webb, born in Tally Ho, North Carolina, whose résumé showed that he was eminently unqualified to run a technical agency.

Jim Webb was not a scientist or an engineer. But he'd been grazed by those people in some of his jobs and knew that they weren't fuzzy thinkers in lab coats or geeks with pencils in their shirt pockets and a slide rule on their belts. He also knew that scientists and engineers shouldn't be turned loose with big budgets and lax supervision.

He'd already met Hugh Dryden, NASA's acting administrator and a longtime engineer and manager, and Webb liked his honesty. But Dryden was prepared to clean out his desk as soon as a new man was named. Webb didn't want the job. He described his own shortcomings bluntly to Kennedy —stressing his background and lack of technical training. Kennedy brushed off his objections. The space race was a matter of high-level policy, not just domestically but on a global scale. Webb could manage the top people who managed the programs, but his real value would be in helping set policy. He'd had policy jobs before, at the Bureau of the Budget and the State Department.

President Kennedy looked Webb in the eye and admitted that he had no space policy, regardless of the Wiesner Report and no matter what he'd said in the campaign or what Webb read in the newspapers. Webb's first assignment would be to propose a policy.

"Take the job," Kennedy said. Webb took the job.

He made Hugh Dryden his deputy and kept Robert C. Seamans, Jr., as associate administrator. Both had strong technical backgrounds. But while Dryden had always focused on programs, Seamans helped to firm up

NASA's administrative structure and to make sure that its headquarters was organized in some kind of coherent fashion.

By keeping Dryden and Seamans, Jim Webb created a bipartisan agency that was at least partly insulated from congressional party politics. In his obituary in 1992, the *Washington Post* quoted Webb in one of its own earlier stories:

"We needed to work together, so here's what I decided: No policy would be approved for NASA until the three of us [Webb, Dryden, and Seamans] had talked it over. None of us would do violence to the strongly held opinions of the other. This was a policy which intentionally put us in chains. We bound ourselves in these hoops of iron. Sure, I could have overruled them. I was the boss. But it wouldn't have worked that way."

Webb also recognized that the space race was a force that could have major impact on education and the economy in general. He decided that every state in the Union would get contracts from NASA, and he made sure that it happened. And he decided that NASA money would not just flow to industrial contractors, but that significant dollars would go to universities through funding their scientific input to space missions and through graduate-level research that NASA could sponsor. He made that happen too.

Most of all, Webb did exactly what President Kennedy asked. He became a major force in establishing JFK's—and thus the country's—space policy in the 1960s.

So Jim Webb took the job for which he was eminently unqualified. And though a few may argue about his methods and because he sometimes proved himself human rather than perfect, he was exactly the right man.

7

The errant Redstone air valve on Ham's MR-2 capsule would have opened anyway when the parachutes deployed. So nobody worried about the water it let in. That little flood came after the holes in Mercury's bottom made the capsule tilt, then capsize, a problem that was quickly fixed by strengthening the heat shield connectors. But the question of why the valve opened during the launch phase had to be answered. It didn't take long. The Redstone rocket had been vibrating like a tight spring as it climbed toward space. The vibrations shook loose the valve's safety pin and it popped open. Now the postmortem focused on those vibrations.

Al Shepard and the rest of the seven weren't too concerned. Nor were others on the manned spacecraft side of Project Mercury. Ham survived his mission in fine physical shape. The medics certified that he'd done his job under all sorts of extreme conditions. It looked like an easy decision. Space Task Group director Robert R. Gilruth talked it over with Max Faget and the rest of his key people, and was ready to declare that MR-3 could launch on March 24 with Alan Shepard aboard.

Wernher von Braun said no.

Politics and cold feet in the Eisenhower administration kept the United States from putting the world's first satellite into orbit. Now politics in the Kennedy administration and cold feet in Alabama threatened to make the United States an also-ran all over again. When the excitement over Ham's flight subsided, a new reality intruded on Al Shepard's dreams. The rocket experts at Wernher von Braun's Marshall Space Flight Center in Huntsville didn't like the vibrations and extra acceleration in the MR-2 Redstone. Ham's ride had been rougher than expected and the trajectory was wrong. No matter that the chimpanzee came through it without a hitch. The idea of putting a man on the next Redstone suddenly seemed too risky.

A week after MR-2, Kurt Debus, one of von Braun's German team and now a key player in launch operations at Cape Canaveral, wrote a note to himself that another unmanned flight was needed. They already were making the changes needed to stop the vibrations, to better control the rocket's thrust, and to fix a short list of other problems, both mechanical and procedural. Their statistical studies of all Jupiter and Redstone flights concluded that the next one's chances of success were between 88 and 98 percent. Bob Gilruth and *his* Redstone experts thought the chances were better than that. They were ready to declare the Redstone man-rated. Von Braun demurred. He reminded his Space Task Group counterparts of Gilruth's own rule. Until everybody agreed, there would be no manned flight.

The Alabama rocket men were making all the fixes, but they wanted one more unmanned flight. The argument between Gilruth and von Braun was bucked up to NASA headquarters in Washington. Jim Webb was brand-new on the job as NASA administrator. He and his just-formed bipartisan team were still getting to know each other. Congress was worried about the Wiesner Report and its implication that the civilian space agency might be soon dominated by the military. And President Kennedy was ordering his own review of NASA and its programs.

If they killed Al Shepard, it would be the end of everything.

The decision took until the end of February, but the pressures were too much. Headquarters wanted to take the safe way out and Gilruth gave in. An unmanned, and unchimped, launch was put into the March 24 slot reserved for MR-3 and Alan Shepard. They called it MR-BD, for "booster development." If it was a success, Shepard could go next.

The date now for MR-3 was April 25, 1961.

The Chief Designer called his cosmonauts together after Chernushka and her robot came home. The little dog was launched March 9, 1961, with an assemblage of wires, tubes, and mechanisms to simulate a man sitting beside her in a Vostok capsule. Sergei Korolev's people had spent more than two months revising Vostok's control and automated retrofire systems. Now they crossed their fingers and hoped that this one wouldn't send the capsule home so steeply that it burned, or boost it into a higher orbit where it would drift dead and dormant for years. The cosmonauts were hoping for the same thing, even more strongly.

After one orbit, Vostok set itself up in the proper position. Its retrorocket fired on time, and everyone knew that a man would soon replace the dog in space. They would do one more test flight, Korolev said, and then one of these six little men around the table would make history. He asked them to do the same thing Bob Gilruth had asked of his astronauts a few months earlier: write down the name of the man, not yourself, who should be chosen. Korolev then went a step further. Gilruth asked only for names. Korolev asked for a few words justifying the choice.

The majority picked the same man. Why? Their reasons were simple, almost childlike. "Never loses heart." "Fit comrade." "Bold and steadfast." The Chief Designer took it under advisement. He wouldn't make his own choice until a few days before it was time to stick a match under an R-7A's tail and send a cosmonaut to glory. Of one kind or the other.

In twenty-six days of March and April 1961, momentous events in technology and battle conspired to confront the United States with an international crisis, and then to change the space race completely.

Mercury-Redstone Booster Development flew a suborbital mission March 24. Bob Gilruth's team gave Wernher von Braun a boilerplate Mercury capsule. They didn't want it back, so it went to the bottom of the Atlantic. All that mattered was the Redstone. It was perfect.

Sergei Korolev's team launched a Vostok the next day. Its passengers were the dog Zvezdochka, "Little Star," and another robotic kludge of pipes, tubes, and instruments. They brought it down after one orbit and it, too, was perfect. On each of the dog missions, successful and not, Tass did not mention the secret Vostok name. The missions, Tass said, were Sputniks.

On a day in early April, the Chief Designer called together his cosmonauts for one final grilling. The next Vostok, he said, was for one of them. He studied each man as he talked, assessed their answers to his questions, and their enthusiasm. He spent rather more time on one of them than the others, the one whose name had come up most often in the peer review. When Korolev left the room, that cosmonaut was certain that he had been chosen. He was right. That evening, April 8, 1961, he and his backup were introduced in a short ceremony to Sergei Korolev's tame journalists. The backup cosmonaut was Gherman Titov.

The man chosen to be first into space was a Red Air Force lieutenant, just twenty-seven years old. His name was Yuri Alekseyevich Gagarin.

Charley Donlan took a promotion and returned to Langley Research Center on April 1. Bob Gilruth looked at his Space Task Group organization and decided that he could do just fine with Walt Williams as his only top assistant. As operations boss, Williams was becoming increasingly powerful and important.

Launch preparations for MR-3, the flight of Alan Shepard, consumed Williams and everyone else assigned to Cape Canaveral. Only a few people knew that it would be Shepard. He was Gilruth's choice, but with Gus Grissom and John Glenn also spending long hours every day in the Mercury trainer and virtually duplicating everything Shepard did, the option was always there for a switch. Their names had been announced to the press more than a month earlier, but not the order in which they'd fly. To most of NASA, and to the rest of the world, who would fly MR-3 was still a mystery. Gilruth wanted it that way to keep unnecessary public attention from pressing down on Shepard's shoulders.

At the same time, a presidential committee was completing its analysis of the Mercury program. Chaired by Donald F. Hornig, a scientist handpicked by science adviser Jerry Wiesner, the committee had been talking to contractors and NASA people, visiting Langley and the Florida launch pads, and generally making Gilruth and his people nervous. The Hornig Report was due at the White House April 12. When a rumor about the report's contents reached Gilruth, he lapsed into uncharacteristic sarcasm. The document, according to the NASA grapevine, would recommend more monkey flights and more centrifuge runs before putting a man into space. It would call for fifty monkeys to be tested, with some of them run to such high g-forces on centrifuges that they died. "If that's true," Gilruth snapped, "we might as well move the program to Africa." There were plenty of monkeys available there.

There was some truth to the rumor. A clutch of doctors, probably including unconvinced flight surgeons speaking to the committee on the sly, had told the Hornig panel that even a few minutes of weightlessness would render an astronaut unconscious. They wanted those additional monkey tests, but whether that would include sacrificing simians on the centrifuge remained a rumor. NASA administrator Jim Webb broke the panel's dead-lock as it wrote its report. "The apes survive zero g," he said. "The monkeys do it." If they could survive, so could Al Shepard.

If the recommendation for additional monkey business was intact on the early morning of April 12, it was gone when the Hornig Report arrived at the White House on schedule a few hours later. Instead, the science panel concluded that putting a man aboard a Redstone was dangerous, but no more so than the dangers faced by aviation pioneers or modern-day test pilots.

An event that came overnight Washington time might have changed some minds.

Yuri Gagarin went to the launch pad shortly after lunch on April 11, with the Chief Designer at his side explaining it all. Sergei Korolev was now fifty-five years old and in failing health. A bad heart and the physical deteriora-tion from prison camp life that had never quite come back to normal left him in a bad mood and tired on too many days. But he was not going to let anyone else show his cosmonaut around on this important day.

Gagarin—blond, healthy, half the age and seemingly half the size, at least in breadth, of the burly, broad-shouldered, dark-haired Korolev—absorbed every detail. He'd be roused from bed at 5:30 a.m. Korolev wanted him to do some stretching and limbering exercises, then eat a light breakfast. The doctors would give him a quick physical exam before he dressed in blue thermal underwear and was helped into his puffy orange space suit. A bus would bring him here, to the launch pad, and sometime around 7:30 a.m., he'd be sealed into the Vostok capsule. Launch was planned for 9 a.m.

His call sign was "Swallow," Korolev said, but his ship would be known as Vostok. The secret name would be secret no more. The Tass news service would announce this first manned flight as Vostok 1, but not until Yuri was safely back on Earth. The Central Committee of the Communist Party had approved Korolev's request to declassify the name. This was no common Sputnik they were planning to launch. History demanded that it have a name of its own. Then in the excitement of the event, the Vostok name wasn't revealed for nearly two weeks. No matter. It was the name *Gagarin* that held everyone's attention.

At some point in the briefing, Korolev committed treason. He gave Yuri Gagarin the three secret numbers that, when used with the three he'd be given as a matter of routine, would let his cosmonaut activate the reentry system. One or two others gave Gagarin the numbers too. Nobody thought that this eager young Russian would defect from space. When they finished, Korolev complained about his heart and led the way to adjoining brown cabins near the control center. He looked tired when he put Gagarin into one, then went next door to rest.

That evening Gherman Titov joined Gagarin. They would sleep this night in the cabin next to Korolev, under the watchful eyes of doctors. Gagarin slept peacefully, without waking. The Chief Designer didn't, checking on his cosmonaut in the middle of the night and certainly thinking ahead to the activities of dawn. He knew, of course, that he was winning. When the United States added that MR-BD launch into its schedule, it handed Sergei Korolev another gift of victory on this lap of the race. He accepted and moved forward to take full advantage of his good fortune.

Yuri Gagarin was too young to be a statesman, but he understood the moment. At the base of the R-7A rocket, he looked at the small crowd of cosmonauts, scientists, and even a few politicians. He nodded at the Chief Designer standing nearby, gathered himself into the best form of attention he could in his bulky orange suit, and said with solemn dignity: "Senior Lieutenant Gagarin is ready for the first flight in the spaceship Vostok."

He began an impromptu Russianesque speech, flowery and patriotic, but stopped and said good-bye when he saw Korolev looking at his watch. A few minutes later he was alone in Vostok. The adventure was about to begin. The countdown delays were minor. Vostok 1 lifted off under clear skies at 9:06 a.m., April 12, 1961, a Wednesday.

"Poyekali!" shouted Yuri Gagarin. *Let's go!*

It was 1 a.m. in Washington, D.C., also Wednesday, and the American intelligence machine was in full operation. They'd known for days that the launch was coming. Jack Kennedy had a moment of trepidation in his box at Sunday's opening game of the Washington Senators, where he'd thrown out the first pitch, when an aide told him that United Press International was ready to announce a Russian manned spaceflight. But it didn't happen that day. Intelligence told the White House to be ready for it overnight Tuesday. Kennedy didn't want to be awakened. "Give me the news in the morning," he said.

So at 1:35 a.m., science adviser Jerry Wiesner roused press secretary Pierre

Salinger from bed to say that the military had tracked a large rocket from the Baikonur-Tyuratam region and that it had put a large object into orbit. They were confident that a man was on board. The White House wouldn't react, Salinger said, until Russia made a formal announcement. A little after 2 a.m., Radio Moscow told the world about Yuri Gagarin.

By then, Gagarin had long ago felt the g-forces pressing him into his couch. It wasn't unbearable. Even above 5 g's, with his lungs and diaphragm compressing, he continued to report to the control center on the radio. When the second and third stages shut down, he felt quick jolts. Then suddenly it was over. There was a moment of queasiness when his body went weightless, seemed to lift up from itself, became a human mass free of gravity. It passed quickly in the excitement. There were things to do and this was to be a very short flight. Before he passed out of Russia, he radioed immortal words to Korolev's mission control.

"I am Swallow!"

His comments mostly were recorded on board. For much of the next ninety minutes, his radio was beyond reach of the few Soviet ground stations or the trawler ships it had deployed. He pulled out a tablet to write down some observations, but the pencil drifted away out of reach. There was much yet to learn about living and working in zero gravity. He clearly saw the oceans and lands below. Over America, traveling north to south in a semipolar orbit, he said that he was in a good mood. On the ground in America, a nation slumbered. Now Yuri Gagarin was orbiting toward home. He'd gotten into space. The next hurdle was to get him back alive. It turned into a close thing. If Radio Moscow had known what was happening up there, it might have waited to tell the world about its first man in space.

Vostok's retrorocket fired on time with the capsule in the correct position for reentry. But when it shut down, something went wrong. Gagarin heard and felt a sharp jolt. He saw through the window more than felt that his craft was spinning wildly. The tumble went quickly to at least thirty degrees per second, a 5-rpm spin that wasn't enough to bring on worrisome g-forces, but enough to force him to concentrate. Africa, the ocean, the horizon, a black sky, suddenly the sun, Africa again passed through his line of sight. The retrorockets hadn't separated. Gagarin said later in his official report that he wasn't afraid. He decided that this was no emergency. Maybe so. Maybe not. The situation was beyond his control anyway.

Vostok spun and tumbled into the upper reaches of the atmosphere, beginning to heat up even in the near-nothingness that it is up there. Yuri

Gagarin could only hold on. It took ten long minutes. Then the retropack let loose. It had come only partially free, trailing behind the spherical Vostok, and the increasing forces and heat of reentry finally took it away. Gagarin was lucky. With the retros gone, Vostok's ball shape sucked the thickening atmosphere around it and the tumbling rates slowed. Still it wobbled back and forth, up and down. He saw fire outside and heard crackling sounds. But the all-encompassing asbestos heat shield was working. Vostok did not come apart. Nor did it flare briefly and brightly and burn man and ship into the nothingness of superheated molecules. The spinning and g-forces grayed out Gagarin's vision for a long moment.

Then it was over. Gravity felt normal. He heard air whistling outside. It was time to bring the adventure to its end, and to begin a deceit that Yuri Gagarin and all the Soviet space experts would play out in public for years to come.

At 23,000 feet, the hatch automatically blew off. Moments later, Gagarin's ejection seat fired him through the opening. All that parachute training was about to become useful. He fell for a few moments, long enough to recognize the Volga River below and to realize that in practice he'd jumped often over exactly this place and that he was familiar with the terrain below. When his parachute opened, he had time to see people in the fields below looking up and pointing. Vostok was below him and off to the side under its own chutes. He settled into a field soft from spring plowing, stood up unsteadily, freed himself from the parachute, and staggered in his bulky orange suit and strange helmet toward a woman and her little girl.

"I'm Soviet," he called. "I've come from outer space."

Anna Takhtarova, local resident of Saratov by the Volga River, approached hesitantly. She'd heard Radio Moscow, but to see this orange-suited man come down under a parachute was overwhelming. "Did you really come from outer space?" she asked.

Yes, he had. The approaching helicopters swooped overhead and landed. A soldier jumped out and called him Major Gagarin. He'd been promoted two ranks during his 108-minute mission into history. Radio Moscow was continuing its broadcast, saying that he had landed safely, calling him *the cosmonaut, Major Yuri Gagarin.*

The White House knew all about it. But nobody had told the right people at NASA. At 4 a.m., a reporter called the astronauts' spokesman for a comment, and Lieutenant Colonel John "Shorty" Powers made one that he regretted when it was widely printed. "We're all asleep down here," Powers grumbled, then woke up enough to congratulate the Russians on their feat, and to keep from saying what he was really thinking.

The damned Russians beat us again!

Al Shepard was livid. He was gracious in his public comments, but in private he raged against the decision to delay his MR-3 flight and against the men, particularly Wernher von Braun, who were too conservative at the one time it really counted. As the years went by, he made his anger and disappointment clear, even to *Life* writers and photographers who came to know him. But he insisted that they not print that story, and they didn't. "What's done is done," he told them, sometimes through gritted teeth.

But much later he went public. In his 1994 book *Moon Shot,* coauthored with Deke Slayton and a couple of veteran Cape Canaveral newsmen, he got fairly blunt about missing his chance to make an American the first man in space.

"We had 'em," he wrote. "We had 'em by the short hairs, and we gave it away."

If he'd known about the Soviet deceit, he might have gone public with sarcasm or anger while Yuri Gagarin was still being canonized in Red Square. No one told the world about Gagarin's personal parachute landing. Soviet space officials promoted the idea that he'd ridden Vostok 1 all the way to the ground. Gagarin himself dodged questions and carefully fudged his replies to press conference questions about his reentry and landing. No one mentioned the nearly out-of-control Vostok and no one mentioned the ejection seat.

The reason was political. The Soviet Union wanted a number of flight records to be made official by the FAI, the Fédération Aéronautique Internationale, in Paris. The FAI controlled the record book for flight and set the rules. So Yuri Gagarin became the record holder, at least until other cosmonauts and astronauts eclipsed him, for the highest flight and for the around-the-world speed record, in addition to being duly recorded as the first man to go into outer space. The problem was, under FAI rules, a pilot must stay with his ship from takeoff to landing. Bail out and it doesn't count.

So some of Gagarin's entries in the scrupulously controlled FAI record book were fraudulent. It was an unnecessary deceit, and one the Soviets admitted in later years. The race was going to the bold and the swift. Yuri Gagarin was the first man in space. Nothing else mattered.

Five days later a ragtag and ill-supported band of rebels invaded Cuba from Florida. First Gagarin, now the Bay of Pigs. As the disaster unfolded, the United States withheld promised air support. Then when the invasion faltered and failed, American troops who had been sitting aboard ships in the Navy port at Jacksonville, Florida, were hastily debarked and sent in threes and fours to Army bases around the country.

No announcement was made that the U.S. Army had been ready to back up the Cubans in their attempt to overthrow the communist dictator Fidel Castro. But when the fourth group of four arrived after midnight a day later at Fort Hood, Texas, the young officer of the day asked why so many fresh-from-basic-training soldiers were traveling from Jacksonville.

"Don't know," was the answer. "We've been sitting on ships wondering what the hell. They just pulled us off and sent us here."

Just as he inherited a shaky space race from Dwight Eisenhower, John F. Kennedy inherited a shaky plan to save Cuba from communism. He got bad advice all around in letting the invasion proceed, and worse advice in abandoning the rebels he'd promised to support. The days of April 12 to April 19, 1961, were not good ones for Jack Kennedy. First Gagarin. Then the Bay of Pigs. He had to do something. The question was what. And when. While he wrestled with the Cuban problem and the harsh media and international reactions to it, he asked his vice president to look at the whole space question. Was there something America could do to eclipse those damned Russians?

Lyndon Johnson, also chairman of the Space Council, took on the assignment, calling in NASA scientists, military experts, and top-level friends in the growing aerospace industry for advice and counsel. Over the same week, all hands at Cape Canaveral got ready for Mercury-Redstone 3, the flight of Alan Shepard. Checkout delays slipped the date from April 25 to May 2, but otherwise it looked good.

Shepard and his backup, John Glenn, spent hours every day in the Mercury trainer, connected to the Cape's mission control center by data and communications lines. As the astronauts reached near-perfection in their reactions to every possible contingency, from a normal suborbital flight to scores of emergencies and foul-ups programmed deliberately into the sessions, so did the men in mission control. Except for one.

Astronaut Scott Carpenter's assignment was the Capcom console. He was the only person who talked directly to Shepard in the Mercury capsule. As the training grew increasingly intense, Carpenter ran afoul of the man called Flight. For a while, Chris Kraft couldn't figure out why some training episodes went so wrong. Then he saw what was happening. At certain points during the Redstone's simulated launch, Carpenter was supposed to activate switches that fed data to the computers. They aborted session after session when the data were simply unreadable. Despite all his preparation, Carpenter couldn't get it right.

"He'd lose track and start flipping switches randomly," Kraft said later, his face muscles still hardening at the memory. Kraft was growing in power, but he was still too junior to throw an astronaut out of the control center. At the end of the day, he called Al Shepard. "Shepard was perfectly reasonable," Kraft remembers, something not hard to be when your life is on the line. The next day Carpenter was shifted to other work and Deke Slayton appeared at the Capcom console.

Shepard blew off tension in the evenings by playing off Bill Dana's routines as Jose Jimenez, the Reluctant Astronaut. "No, no," Shepard would cry in a passable imitation of Jose's Latino accent, "don't do this to me." If he sounded too serious, John Glenn was ready to swap seats.

There was another halfhearted attempt to scrub it all in Washington. White House science adviser Jerry Wiesner was having second thoughts about the aftereffects of a failure. The nation was literally holding its breath in anticipation of an American in space. NASA agreed to live television coverage of the launch, beginning at T minus two minutes. It was a first, but it was a way to upstage the Russians. Amid the acclaim for Yuri Gagarin's spaceflight were loud growls that it was all kept secret until after he was in orbit. NASA was a civilian agency and Mercury was an open program. From now on, television coverage of important flights would be live. If disaster struck, the whole world would see it in real time. JFK himself was briefed on the plans. Paul Haney took a call from Evelyn Lincoln, JFK's secretary, and then talked to Pierre Salinger, Kennedy's press secretary, to assure them about the abort system's capabilities and testing. Kennedy did not cancel the television coverage.

Finally May 2 arrived. The lucky number 7 was everywhere. Mercury capsule number 7 sat on the launch pad atop Redstone rocket number 7. There were seven astronauts, and Bob Gilruth had given them permission to decide on their own call signs. Al Shepard thought it over and picked *Freedom 7.*

The countdown started a day earlier. Then a weather front moved into the Cape area. Walt Williams ordered the launch scrubbed. It would take forty-eight hours to recycle. They'd start again on May 4 and try to launch on May 5. After the scrub, reporters finally figured out who was going to be the first American in space. Shepard was the one suited up and ready before the weather got bad. NASA acknowledged that he was it, and his name flashed around the world.

They'd roused Yuri Gagarin from his bed only three and a half hours before launch. Al Shepard wasn't so lucky. He was up at 1:10 a.m. and went through more than five hours of preparation. He had steak and eggs for his middle-of-the-night breakfast, endured yet another physical exam, and had biosensors glued to his body. They did a few more Jose Jimenez bits as the hours wore on. By 3:30 a.m., he was being helped into his space suit. Launch was set for 7:20 a.m.

Life's Ralph Morse caught Shepard in crisp black-and-white photos getting out of the van that brought him to the pad, holding his portable air-conditioning unit, and grinning as he entered the elevator that took him to the top of the tower. Getting into the cramped Mercury capsule wasn't easy; eventually his suit hoses were hooked up, the air conditioner was removed, and with a finality that sent his heart racing, the hatch was pushed into place and bolted shut.

Al Shepard was alone with his destiny. The feeling didn't last long. Astronaut Gordo Cooper chatted him up from the blockhouse that controlled launch. Then they called a hold when dawn clouds rolled across the Cape. It took fifty-two minutes for the weather to clear. Shepard had been sealed in his space suit for more than four hours now. He felt the pressure in his bladder increasing. On the original schedule, he should already be bobbing out there in the Atlantic waiting for the recovery helicopter to pick him up. Instead he was on his back in a sealed capsule with an hour or more to go before somebody pushed the Redstone's Go button.

Once the bladder distress reached the conscious level, it quickly got worse. Shepard asked if he could get out and hit the head. Cooper got a quick answer from the blockhouse experts. *No.*

That just intensified the problem. Shepard thought it over and made a commander's decision. He'd simply urinate in place. His thermal underwear would soak up the moisture, and the steady flow of pressurized, almost humidity-free oxygen pumping through his suit should dry everything out before launch. He didn't ask for permission, just radioed his decision to

Cooper and did it. It was decidedly uncomfortable, but not for long. The countdown resumed, Shepard's mind was occupied, and by the time he thought of it again, he was completely dry.

There was another short hold for computer problems, and then at 9:32 a.m. communications switched from Gordo Cooper in the blockhouse to Deke Slayton in the control center. At the same time, the three television networks went live with their coverage, and 45 million Americans leaned forward to watch black-and-white images of a black-and-white Redstone rocket on the pad and ready to go.

The blockhouse countdown went toward zero.

Flight demanded a status from each console in mission control. Deke Slayton listened and as each "go" came over the loop, he relayed it to Shepard. The only transmissions Shepard remembered clearly were the last seconds of the countdown and Slayton's cryptic call as the Redstone lifted off.

Wally Schirra put an F-106 into a steep bank just above the wave tops and a mile offshore. Scott Carpenter circled in another F-106 at 31,000 feet. The sky had cleared to a crystal blue and Schirra watched the Redstone. His job was to put the 106 into an afterburner climb, turn on his cameras, and follow the accelerating Redstone as long as he could. When he couldn't keep up anymore, Carpenter would take over and climb to the limits of his aircraft.

". . . two, one, zero," was the call from the blockhouse.

Deke Slayton waited a beat, made sure that the Redstone was climbing away from the pad, pressed the key to his microphone, and transmitted the first words heard by an American on his way to outer space.

"You're on your way, Jose."

There was no reluctant astronaut aboard *Freedom 7* that day.

"Roger. Liftoff and the clock has started," Shepard's crisp voice responded. The national reaction was aggressively euphoric. Except in the immediate vicinity of Cocoa Beach, Florida, Americans threw their arms in the air, cheered and yelled at their black-and-white television sets, and stopped yelling "Go! Go! Go!" only long enough to wipe tears from their cheeks.

America's runner in the space race was Navy Commander Alan B. Shepard and he was on his way. A nation's pride and dreams and hopes ran skyward with him. They didn't watch television in Cocoa Beach. There is no better description of what happened there than this one from *Moon Shot:*

> In Cocoa Beach, people left their homes to stand outside and look toward the Cape. They went to balconies and front lawns and back lawns. They stood atop cars and trucks and rooftops. They left their morning coffee and bacon and eggs in restaurants to walk outside on the street or on the sands of the beach. They left beauty parlors and barber shops with sheets around their bodies. Policemen stopped their cars and stood outside, the better to see and hear. Along the water, the surfers ceased their pursuit of the waves and stood, transfixed, swept up in the fleeting moments.

High and long and to the east, Al Shepard and *Freedom 7* disappeared into the sky. A contrail from Scott Carpenter's F-106 looped over where it reached the jet fighter's maximum altitude, far below the place where Mercury was going. From the blockhouse, Shorty Powers, the man who soon was known as the Voice of the Astronauts, or the Voice of Mercury, was on the air telling reporters at the nearby press site and the rest of the world through live feeds to radio and television networks that everything was "A-OK." It was a term from the old days of the telegraph, easy to send in Morse code. Now it entered the American lexicon and became the phrase that instantly told the world out there that everything was fine. Lieutenant Colonel Powers used the "A-OK" ten times in the next minutes and much more in missions ahead. No astronaut ever used the phrase and some bristled when headline writers attributed it to them. "That's not a pilot's term," they'd say. "It was Shorty."

MR-3 was a near-textbook mission. *Freedom 7* came free of the burned-out Redstone rocket, the escape tower flipped off, and Shepard felt the stomach-sinking rush of weightlessness. It was not unpleasant and he could see. He could move. He could think. He could talk and he radioed his reports on the capsule's condition one after the other to Deke Slayton in mission control. His coordination was fine. Reaching for switches, turning knobs, gripping and twisting the hand controller that let him change the capsule's orientation—no problem, not a bit. The damned doctors had been wrong all along and now he had the proof.

He switched to manual control and marked up a first in space for America. He did change the way *Freedom 7* pointed, little hydrogen peroxide thrusters responding to his hand's input and moving the capsule's nose this way, then that way. Yuri Gagarin was a passenger. Al Shepard was a pilot.

He could see all the way north to the Carolinas, west to Mobile Bay, south and east along Florida and to the sprinkling of brown Bahamas set into turquoise patches of ocean. The higher *Freedom 7* climbed toward its apogee of 116 miles, the more he could see. In no time at all he was over the top, inertia gone, and falling back toward Earth. Shepard set up the capsule's position and test-fired the retrorockets. He didn't need them to get home, but somebody else would. It was nice to know that they worked.

Down he dropped through the atmosphere until he began to worry about the capsule's parachutes. But the little drogue chute popped out at 21,000 feet, soon dragged the sixty-three-foot main parachute from its canister, and at something below 10,000 feet it, too, opened. Three minutes later *Freedom 7* splashed down, only 303 miles from Cocoa Beach. The capsule tilted over briefly, then turned upright. It was no worse than the thump of landing a jet on a carrier deck, and that was a feeling Al Shepard knew with long familiarity. Almost immediately a helicopter was overhead, hooking up a recovery cable and beginning to lift the capsule from the water. Shepard did his shutdown chores, opened the mechanical hatch latches, and climbed out of *Freedom 7* and into the horse collar man lift. Moments later he was aboard the chopper, and just eleven minutes after splashdown, he stepped onto the deck of the aircraft carrier *Lake Champlain*.

The little boy in Al Shepard couldn't suppress either his elation or a quick joke. Before returning the salutes being offered, he looked at the officers and men, took on exactly the right accent, and said, "My name Jose Jimenez."

They flew him to Grand Bahama Island for debriefing. But first he got an unexpected call from President Kennedy, who had watched television with the rest of the nation. The roar of approval for Al Shepard already was being heard in every corner of America. Kennedy felt the heroic tumult building, but even his fine-tuned politician's antennae didn't foresee its mass. He was simply being president and congratulating a hero.

Al Shepard was a hero, but he was still Al Shepard. During the debriefing sessions, he was brilliant. He remembered everything. To debriefers' questions, he gave detailed and expansive answers. He'd only been weightless for five minutes, but his observations were burned into his memory. He offered insights. He made recommendations. He was as professional as an engineer and a test pilot could be. He was, Walt Williams would say, awesome. Then around three in the afternoon, a pretty young secretary brought coffee into the session. "You could see Shepard's brain get up, leave the room, and follow her down the hall," Williams laughed. "He wasn't worth a shit the rest of the afternoon."

So heroes aren't perfect. But usually they're forgiven, especially for lapses that aren't widely known until decades later. If the nation had heard that story the day it happened, it wouldn't have cared. Through the next week, in a crowded parade down Pennsylvania Avenue for a hero with a stunned look on his face, at a White House reception, on the covers of magazines and the front pages of newspapers, over and over again on television, Al Shepard was hailed and the American people rose up to throw their hearts and souls into the space race against the Russians.

Jack Kennedy absorbed it all and asked his vice president to finish up his report and make a recommendation. The mood was right and the country was ready. Kennedy already knew which way Lyndon Johnson was leaning. He'd mentioned it to Al Shepard that day at the White House and sworn him to secrecy. He also talked to Bob Gilruth. What could they do in space that the Russians couldn't do? "Well," Gilruth told him, "you've got to pick a job that's so difficult, that's new, so they'll have to start from scratch. They just can't take their old rocket and put another gimmick on it and do something we can't do."

With that comment, Gilruth put his finger on the core of the Chief Designer's spectacular successes: He had a big rocket, but not much else. Russian spaceflights were gimmicks. Whatever Kennedy decided, Gilruth told him, "It's got to be something that requires a great big rocket, like going to the moon. Going to the moon will take a new rocket and new technology." When Kennedy looked doubtful, Gilruth pressed on. "If you want to do that, I think our country could probably win because we'd both have to start from scratch."

JFK was hearing the same thing from Lyndon Johnson and his advisers. He hardly needed to think it over. That was just what Bob Gilruth was hoping. "He was a young man," Gilruth said much later. "He didn't have all the wisdom he would have had. If he'd been older, he probably never would have done it."

But he did. It would resurrect his presidency, and that wasn't all. The stakes in the competition with international communism were too high to play with a weak hand. It was time to take the race to a higher level. President John F. Kennedy spoke to Congress and to the country on May 25, 1961. He talked honestly, bluntly, from his still youthful and idealistic heart. One sentence of his speech would be engraved on the gates of history:

"I believe this nation should commit itself to achieving the goal, before this decade is out, of landing a man on the moon and returning him safely to Earth."

Some of the most famous photos ever taken in space showed a distant planet Earth rising above the lunar horizon during the Apollo 8 mission on Christmas Eve, 1968. *(NASA photo)*

Apollo moon astronauts saw a full Earth, a blue and white and brown pebble against a black sky. *(NASA photo)*

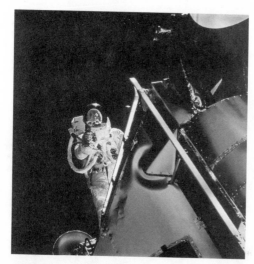

Astronaut Rusty Schweickart was first to exit lunar module and stand on its porch during Apollo 9, which stayed in orbit around Earth to test the strange-looking ship. *(NASA photo)*

Artist's concept shows the Apollo command and service modules docked with lunar module. *(NASA photo)*

Saturn 5 on pad. A test version of the mammoth Saturn 5 moon rocket, 363 feet tall, was rolled out to the launch pad in 1966. *(NASA photo)*

NASA hoped that lunar bases would be a natural follow-on once men landed on the moon. *(NASA photo)*

Buzz Aldrin took no photos of Neil Armstrong on the lunar surface.
This view of Aldrin standing next to the flag was taken by Neil Armstrong.
(NASA photo)

A Titan rocket blasts off carrying the Gemini IX-A spacecraft with Tom Stafford and Gene Cernan aboard. *(NASA photo)*

Gene Cernan became so exhausted on the Gemini IX-A space walk that his time outside the spacecraft was cut short. *(NASA photo)*

Every Gemini mission splashed down safely, most of them within sight of the recovery forces waiting to pick the astronauts up. *(NASA photo)*

Nose-to-nose Gemini VII. Wally Schirra pulled Gemini VI to within inches of Gemini VII's nose in December 1965. *(NASA photo)*

The two spacecraft are 210 feet apart.

Distance between two spacecraft is 55 feet.

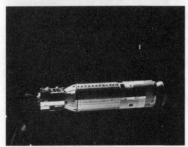

Distance between two spacecraft is 44 feet.

Distance between Gemini VIII and Agena spacecraft is two feet.

Gemini VIII Agena views. Shortly after docking with their Agena target in March 1966, Neil Armstrong and Dave Scott experienced one of the most dangerous emergencies in U.S. space history. *(NASA photo)*

Astronaut Ed White became America's first spacewalker in 1965. Cosmonaut Aleksei Leonov's space walk a few months earlier almost ended in tragedy. *(NASA photo)*

Neil Armstrong and Buzz Aldrin train for exploring the moon in this photo taken by Ralph Morse for *Life* magazine and NASA.

Apollo patches.

Cernan with Snoopy. The Apollo 10 astronaut brought Snoopy to a press conference. The lunar module he and Tom Stafford flew to within 50,000 feet of the moon was named for the cartoon dog. *(NASA photo)*

The author (front row, fourth from the left) at a Gemini press conference in Houston, 1965. *(NASA photo)*

Reporters gathered around Neil Armstrong at a Gemini VIII press conference in 1966. Author Schefter is partially visible at far right. *(NASA photo)*

Astronaut John Young caught a quick view of the module *Snoopy*, carrying Tom Stafford and Gene Cernan, as it separated from the command module *Charlie Brown* in May 1969.
(NASA photo)

The command module *Charlie Brown* was a welcome sight to astronauts Tom Stafford and Gene Cernan as they completed their first rendezvous in lunar orbit, May 1969. *(NASA photo)*

Snoopy ascent stage. Only the stage remained of the lunar module *Snoopy* when it rendezvoused with the command module *Charlie Brown* in lunar orbit, May 1969. *(NASA photo)*

Author Jim Schefter at mission control console in 1965, where he spent days
learning intricate details of upcoming space missions. *(NASA photo)*

Ham, the space chimpanzee, was wired up and ready for his flight into space in 1961. *(NASA photo)*

Wally Schirra clowned with Jose Jimenez (comic Bill Dana, at top) in a Gemini spacecraft in 1964. Dana was a close friend and confidant of the astronauts. *(Mickey Kapp)*

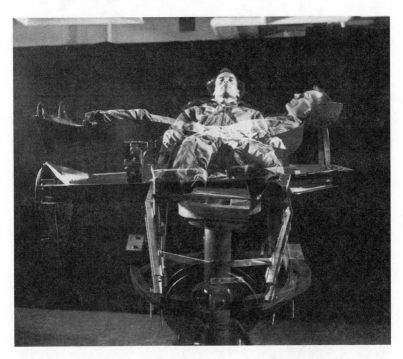

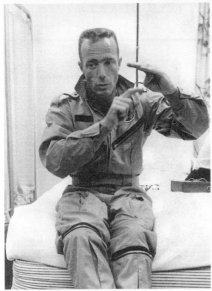

Astronaut Deke Slayton's response to spinning and tilting was part of his training for a Mercury mission. A minor heart problem later grounded him.
(NASA photo)

Scott Carpenter discussed the sensations of weightlessness with doctors immediately after his three-orbit Mercury mission. *(NASA photo)*

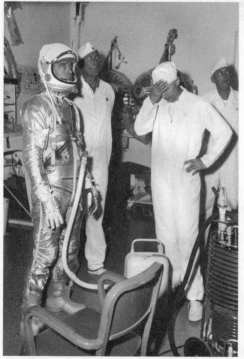

The final Mercury
mission safely ended,
three high NASA officials
sat back in mission control
at Cape Canaveral.
Left to right:
Dr. Robert R. Gilruth,
D. Brainerd Holms,
and Walt Williams.
(NASA photo)

Alan Shepard waited with
launch pad technicians
for word that he could
board *Freedom 7* for his
historic moments as
America's first man in
space. *(NASA photo)*

8

Two weeks after his momentous declaration, Jack Kennedy met Nikita Khrushchev for face-to-face talks in Vienna. Over lunch, Kennedy suggested that the two countries cooperate on a lunar program. Khrushchev said "no," "maybe," then after returning to Moscow, made it a definite no.

Political analysts and newspaper columnists the world over criticized the Soviet Union for its negative position. The most-asked question was "What do they have to hide?" Sergei Khrushchev asked his father about the decision during one of their frequent after-dinner walks around the Kremlin grounds.

"If we cooperate with the Americans," the elder Khrushchev said, "it will mean opening up our rocket program to them. We have only two hundred missiles, but they think we have many more."

Personal misunderstanding and distrust ran so deep between the United States and the Soviet Union that Khrushchev was convinced Kennedy would attack tomorrow morning if he discovered that fact. Soviet rocket force weakness was masked by the Chief Designer's successes in space and by Nikita Khrushchev's saber-rattling bluster. "So when they say we have something to hide . . . ?" Sergei asked.

At that, his father laughed. "It is just the opposite. We have nothing to hide. We have nothing. And we must hide it."

Until Jack Kennedy reached for the moon to atone for the early lapses in his administration, the space race was based on one-upmanship. Beating the other guy was national policy in both the United States and the Soviet

Union. But it was like some orbital poker game. *I'll see you and raise you one.* And there was always another hand. Jack Kennedy redefined the race. Now there was only one pot and its name was the moon. Winner take all.

There was only one kind of spacecraft that mattered too, and it was manned. Enormous strides were coming in satellites of all varieties—weather, communications, scientific, spy—and the technological and industrial might of the United States would take command and quickly leave the Russians far behind. In short order, America dominated outer space and Americans were first to reap the benefits. But that part of the space race went off to a dead-end siding in the public mind. The only race now was to the moon. It was a finish line everyone could see from the backyard. There was a time limit too: the end of the decade. And it had to be a man who got there and came back alive. Robots and machines didn't count.

While the Chief Designer digested the new rules and got ready for his next mission, America went back into space. This time it was Gus Grissom on Mercury-Redstone 4, the last of the manned Redstone flights. His capsule was *Liberty Bell 7,* and what happened gave everyone in the Mercury program a sinking feeling. Grissom's job was to duplicate the Shepard flight. With two successful manned flights into suborbital space, Bob Gilruth and the Space Task Group would consider declaring Mercury ready for orbit. Then they'd do a monkey flight to orbit before clearing the way for John Glenn to get his ride. Al Shepard was at the Capcom console when Grissom lifted off on time at 7:20 a.m., July 21, 1961. Only two other men had ridden a space rocket, and Shepard did his bit to lighten the moment.

"Jose, don't cry too much," he radioed to Grissom just eight seconds into the flight.

Grissom choked back a laugh and responded with the standard Jose Jimenez line: "Oke-doke."

Twenty minutes later he was choking on seawater, not laughter.

The flight was flawless. Grissom and *Liberty Bell 7* soared to 118 miles, logged nearly six minutes of weightlessness, and splashed almost exactly where Shepard had, 302 miles from the launch pad. His capsule had one big change from Shepard's. The recovery people decided that the mechanically latched hatch was too hard to remove from the outside. In an emergency at sea, thermal shingles and bolts had to be removed to get it free. The new hatch was surrounded with explosive primer cord. When the cord blew, the hatch popped free. The recovery helicopter was turning toward *Liberty Bell*

7, its pilot talking to Grissom. "When you blow the hatch, the collar will already be down there waiting for you . . ."

"Ah, roger," Grissom replied.

Hardly a second later the hatch blew. Water began pouring in. Grissom didn't hesitate. He grabbed the edge of the instrument panel and bailed. He was briefly underwater, then sputtered under the heavy downdraft of helicopter blades directly overhead. He began to drift away. His space suit had a rubber neck dam designed to keep water out and verified in tests by Wally Schirra. The dam probably saved Grissom's life. By now, *Liberty Bell 7* was sinking. One chopper hooked onto its metal recovery loop with a shepherd's crook and cable and attempted to lift it from the water. Another tried to drop the horse collar to Grissom, but he was moving away in the downdraft and he, too, was sinking. Water was pouring into his suit through an open hose vent.

"I reached down and locked [the vent] and that stopped any more water from coming in the suit," Grissom told his debriefers later. "Swimming around in the suit was pretty difficult, always trying to keep my head up out of the water."

The official transcript didn't mention the heavy hoard of souvenir dimes in his boots and space suit pockets. He took them along to give to friends and family. Now they were dragging him down. Then he got hold of the horse collar, slipped his head and arms backward into its salvation, and was lifted up to the chopper.

Liberty Bell 7 didn't make it. Films show a helicopter struggling to stay in the air, almost saving the capsule, then losing the fight as engine bearings burned and more seawater sloshed through the open hatch. Just before the chopper would have been pulled down too, the crew chief cut *Liberty Bell* loose. It sank in seven thousand feet of water.

The press blamed Grissom for losing his ship. The inevitable leaks to reporters said that he'd blown the hatch early, either deliberately or by accident. Grissom vehemently denied it, and the other astronauts backed him all the way. "I took the detonator cap off and put it down toward my feet," he said in his debriefing. "Then I pulled the safety pin out and I dropped it, as I recall. I was just lying there, waiting for the chopper to come in . . . I heard this 'pow' or dull thud . . . I looked up and could see blue sky, and water running into the capsule."

Investigators were convinced that Grissom had accidentally hit the firing plunger with his shoulder as he wiggled around, or had gotten so excited after his quick spaceflight that he automatically removed the cap, pulled the safety pin, and blew it, all in one continuous, highly trained reflex. No matter what they tried in the laboratory, they couldn't make similar units just blow by themselves.

Wally Schirra spent hours in a trainer, fully space-suited, trying to hit the plunger with his shoulder. He couldn't do it. While the press continued to roast Grissom, and he pointedly was not invited to a warm reception in Washington, the other six astronauts rallied round. The official report didn't quite exonerate Grissom, but it didn't condemn him outright either. In carefully worded findings, it said that no reason except for accidental activation by an astronaut could explain why the hatch blew prematurely. It also carefully said that such an accident could not be proved. Between the lines, the report was clear. *Grissom did it.*

But lacking absolute proof, he couldn't be censured. The most lasting effect of the near-disaster was Grissom's magnified intolerance for the press. It would be years before he could look at a reporter without wanting to punch him out. There was a footnote. Four more astronauts flew in the Mercury program. After recovery, each of them had a bruised hand from hitting the hatch plunger. Gus Grissom's hand wasn't bruised.

The race had changed, but running in it was still a step-by-step process. The Space Task Group hadn't put a monkey or a man into orbit yet and now Gherman Titov was up there in Vostok 2 barely two weeks after Gus Grissom's sinker, not just staying in space for orbit after orbit but sound asleep to boot.

American euphoria vanished again. The carping from Congress and the news media returned. It was nearly three months since Kennedy declared that the moon was the finish line in the space race. And yet NASA didn't have plans for a lunar vehicle and hadn't begun to master the techniques of manned spaceflight. At Marshall Space Flight Center, Wernher von Braun's group was getting ready to test a giant U.S. rocket, the Saturn 1, with an F-1 engine that would have 1.3 million pounds of thrust. An even bigger Saturn with clusters of F-1s, maybe as many as five, in its first stage, was being discussed. But nobody knew whether such a rocket was moon-capable. And there was talk of an even bigger rocket that might be called Nova. But the multiengine Saturn would be years off, and Nova was speculation.

Titov came home after a full day of weightless space travel, completing seventeen orbits of Earth, sleeping for nearly seven hours, and doing the same kind of spacecraft attitude control demonstrated by Shepard and Grissom. He also reported a phenomenon that almost every space flier would experience, a nauseous feeling like seasickness. It was caused by his inner ear trying to find equilibrium in zero g, and it would pass. Titov didn't vomit; others to this day are not so lucky.

Years later, Soviet space experts revealed that Nikita Khrushchev himself set the date for Titov's flight, August 6, 1961. He was now using space as a full-strength political and propaganda weapon. Just as the international acclaim for Titov reached its peak, Khrushchev gave the nod to East Germany. That day they began building the Berlin Wall. In the face of the Soviet Union's demonstrated mastery of space, and its vaunted (though now quietly questioned) lead in military missiles, the West did nothing but lodge complaints. It was the third blow in international relations to Jack Kennedy in just seven months in office—first Gagarin, then the Bay of Pigs, and now the Wall. JFK was not having a very good year.

With Titov's flight and his safe return, the Chief Designer's Vostok program entered a strangely quiet phase. Sergei Korolev had temporarily exhausted his program's capabilities. There were no great "firsts" to be had with Vostok, at least none that would even remotely contribute to a moon race. A new spacecraft was on the drawing boards, one with maneuvering abilities far beyond Vostok and big enough to carry several crew members. But this Soyuz ship couldn't be rushed into production like the first Sputniks and Vostoks. When Kennedy upped the ante in the space race, he unknowingly forced Sergei Korolev to take a slower, more methodical approach to the competition.

Mercury was slowing down a bit too. After two manned suborbital flights in two and a half months, it was time to retire the aging Redstone rocket and get on with flights powered by the Atlas booster. Redstone retired all right, but production and checkout problems on both the Atlas and the new Mercury capsules on the assembly line put a hitch in the Mercury plans. While Gherman Titov was still catching his breath after landing, NASA said that it would be January 1962 before it could send an astronaut into orbit.

There were other chores taking up time too. Bob Gilruth's Space Task Group had outgrown facilities at Langley, Virginia, and needed a new home. For the first half of 1961, NASA had been surveying sites for Gilruth's operation. The new space center would be home to the astronauts and much of their training, to a new and advanced mission control center, and to the designers working on moon craft. Florida politicians and businessmen howled at the idea of losing the control center. But in fact, once the rocket and spacecraft disappeared over the far Atlantic horizon, the control center's location didn't matter. It could be anywhere. Like Houston.

Politics naturally played a role in picking the new space center's location. Texan Lyndon Johnson was vice president of the United States and Houstonian Albert Thomas was chairman of a key House appropriations subcommittee. Along with high-powered Houstonians including Oveta Culp Hobby, owner and publisher of the *Houston Post,* they'd pulled a classic Texas land deal. First, Humble Oil Company (later to become Exxon) donated a big parcel of prairie and wetlands on Clear Lake, about twenty-seven miles south of Houston, to Rice University. The land overlay a vast reservoir of oil and natural gas and was populated by cattle, red wolves, prairie hens, waterfowl, and snakes. Humble retained the mineral rights.

After holding title just twenty-four hours, Rice donated 1,000 acres to NASA and agreed to sell it another 650 acres at $1,000 an acre. Rice retained some land, including an empty and decrepit mansion adjoining the NASA property. Humble got a huge tax deduction without giving up what it really wanted, the oil and gas below. Rice made some money, still had some land left, and assured itself of close ties with a major NASA center. NASA got free and cheap land. Only the taxpayers, who saw Humble's tax payments go down slightly for a year, took a hit. But they got it back in spades when the land around the new center was developed and populated.

The politicking aside, Houston met NASA's basic criteria, which included proximity to major universities, a nearby international airport, major transportation networks, a local technical infrastructure, and more. Its marks were as high as other sites and higher than most—including Boston, Cape Canaveral, Tampa, and New Orleans—and the politics were right. Houston was chosen. But just before NASA was set to make the announcement, Hurricane Carla struck the area a devastating blow. NASA held off until the cleanup was well under way.

The weather was fine in Florida. They launched Mercury-Atlas 4 with a man-simulating robot on board to verify the environmental system during a one-orbit flight. There was a problem with the oxygen system, but Mercury switched to a backup and could have done eight full orbits with an astronaut on board. Still, they declared this one only a "partial success" and set about to find the source of the oxygen problem. It was a loose handle that could have been adjusted by an astronaut. Just to be safe, they decided to design a new handle.

With the mission completed, Bob Gilruth boarded a NASA airplane with a few of his staff. Paul Haney from public affairs was along, hitching a ride back to Washington. But once in the air, Gilruth had the pilots divert nine hundred miles west to Houston. He wanted to see the new space center site for himself.

"It was all water down there," Haney said. "A shrimp boat had even been blown in by the hurricane." He asked if Gilruth was planning to build his space center on an ocean platform.

"It's only a few feet deep," Gilruth grinned. "It'll dry out. That'll be dry land." He was partly right. The water disappeared, and by the next summer, construction of the new Manned Spacecraft Center was beginning. But as Gilruth would discover when he moved from Virginia, there are only two kinds of weather in Houston—hot and humid, and cool and humid. Houston never dries out.

The three-man spacecraft they'd been calling Apollo for the last year— something to do after Mercury, maybe even a flight around the moon and back—opened eyes and minds to new complexities. When Jack Kennedy made *landing* a man on the moon both a goal with a deadline and a national policy, the complexities got a whole lot worse.

When they sent the first tentative missions toward the moon, both the Russians and the Americans used the "direct ascent" method. They'd launch a rocket at the right time, aimed at the right point (they hoped) to intercept the moon, and the rocket would simply keep firing its stages until the payload broke free of Earth's gravity and was in translunar space. The Russian probes could be bigger because their rockets were bigger. But neither country had a rocket big enough to send a manned spacecraft on a direct ascent to the moon. Both were developing bigger rockets. But they knew that it could take a rocket with 11 million pounds of thrust or more to do a lunar landing. The mythical U.S. Nova might be that big. But only the Saturn 1, with 1.3 million pounds of thrust, was actually being built. On the Russian side, too, there was nothing in the works approaching double-digit thrust levels.

There was another scheme that could use smaller rockets. It was called Earth orbit rendezvous, or EOR. *(The Book of Acronyms* at NASA grew daily. The day would come when listening to aerospace people talk, even at barbecues and parties, was like listening to a couple of Greeks conversing in Swahili. Strangely enough, it didn't take long before wives, kids, and space reporters spoke the language too. For reporters, the problem was to translate it back into English.) With the EOR method, two or three smaller rockets would launch spacecraft into orbit. These ships would rendezvous, dock together, then fire their own much smaller rocket to get to the moon. The combination would go directly to the moon and land, like hitting a bull's-eye target. Finally one piece of the combination with a much, much smaller rocket would blast away from the moon with crewmen aboard and zoom directly home to Earth. The problem was rendezvous. Nobody had ever done it, and the laws of physics made it extremely tricky. A Mercury capsule couldn't do it at all. It lacked the onboard rocket power to make sideways adjustments in its orbit and couldn't carry enough fuel to do the fine-tuned maneuvering required in the last stages of rendezvous and docking.

Enter Mercury Mark II. Bob Gilruth had given Max Faget an assignment to study a two-man spacecraft that could stay in orbit more than a day or two and had the maneuvering power to rendezvous with another ship. Kennedy's moon speech gave Faget and his Flight Systems Division more reason to hurry. The end-of-the-decade deadline almost certainly meant that EOR would be the method. Even the wizard von Braun couldn't do a rocket big enough for direct ascent in the time allotted.

By midsummer of 1961, Faget's team had an enlarged Mercury capsule on the drawing boards, with service and propulsion modules attached at the bottom. It was too big for an Atlas rocket, but the Air Force had a new ICBM called the Titan II that could do the job. And an Air Force upper-stage rocket named Agena just might make the perfect target vehicle for an orbital rendezvous.

Gus Grissom still hadn't flown in Mercury when Gilruth sent the concepts on to NASA headquarters. Mercury Mark II could be the perfect interim space program between Mercury and Apollo. And Apollo itself was being revised at the same time. The three-man ship sketched more than a year earlier by Faget's people couldn't land on the moon. They needed some kind of lunar excursion module to do that job.

By the time they picked Houston as home to the new Manned Spacecraft

Center, NASA was circulating the Mercury Mark II plans to the now-booming aerospace industry and was in preliminary talks with the Air Force about Titan II and Agena. Mercury Mark II was too much of a mouthful, so somebody suggested a new name. The spacecraft would carry two men, twins in space.

Why not call it Gemini?

Enos the chimp made a memorably perverted kind of history in the aerospace subculture, gave the man called Flight a big test of his mission control mastery, and cleared the way for the making of an American hero. There were the usual nit-picking problems and delays before the countdown for the last monkey flight hit zero on November 29, 1961.

The Russians had been quiet in space since Gherman Titov's seventeen-orbit flight in August. Now nearly four months later, the Americans were getting back into the race, but with a monkey, not a man. Bob Gilruth took heat in the press and Congress for not going straight to a manned flight, but he stuck to his decision and Jim Webb backed him. The Mercury equipment needed one last checkout in orbit, Gilruth said, before he'd risk sending a man up there.

Enos got the job because of his proven skills at doing the right thing under pressure. His method of releasing pressure caused some problems, though. He'd pull down his diaper in the middle of a training exercise and begin to masturbate. His handlers and the medics figured that he'd stop if they inserted a catheter to drain off urine instead of using a condomlike device attached to a tube. It didn't work. When he got the urge, Enos would yank out the catheter and go to it.

They devised an advanced catheter with a small inflatable balloon to prevent its easy removal. That worked up to a point. Unfortunately that point was in outer space.

What happened to Enos in orbit proved that Gilruth had it right. They needed one more trial in space. The medics had devised four work tests for Enos. First came the same test given to Ham: pull the right lever when a light flashes and avoid a shock to the foot. In the next test, Enos had to wait twenty seconds after a green light came on before activating a lever. If he did it right, he got a drink of water. If he did it wrong, the test started over. Test three gave him another reward; after pulling a lever fifty times on cue, he got a banana pellet. And in the fourth test, three symbols would appear over levers—two circles and a square, or two triangles and a circle, maybe two squares and a triangle. The challenge was to pull the lever under the odd symbol. If he did it wrong, his left big toe got shocked.

That was the one that made him mad.

After one turn around the world, everything looked good. In the control center, flight director Chris Kraft polled his men, and the recommendation was unanimous: "We're go, Flight."

Two more orbits and they'd bring Enos down. Then it began. An electrical unit overheated. The environmental control system went wacky. The temperature in Enos' sealed couch unit started going up. And Enos himself was one mad monkey. He'd been doing his jobs fairly well. But on that fourth test, a switch short-circuited. No matter what lever Enos pulled, even the right one, his foot got zapped. He kept working and the unit kept hitting him with shocks. Added to his rising body temperature, Enos was faced with a sweaty ride and a test with no right answer.

Then a thruster failed. They'd find debris in its fuel line when they tore it down back home. But for now, the absent thruster meant that the Mercury capsule was no longer stable. It was tilting sideways. When it rolled thirty degrees, another set of thrusters would ping it back upright. Then it tilted again. The process was using fuel at an alarming rate.

The man called Flight demanded explanations and recommendations. He got a "go" from the medics; Enos was stabilized at 100.5 degrees. He got a "go" from the environmental people; a heat exchanger had iced up, then started working again. He got equivocation from the systems people; they wanted more time to analyze the thruster problem. Enos was passing Australia, heading toward Hawaii. Flight gave them their time. But while the experts looked at fuel usage and possible reentry problems, Flight warned Hawaii to get ready for new orders. They might have to send signals to the capsule to come down early.

Out there in space, Enos kept trying to find the right lever. No matter what he did, his reward was a zap to his toe. Banging on the control panel didn't help, so he almost wrecked it. Trying to throw banana pellets in the cramped couch was no good. Ripping through his pants only made a mess. But somehow he got a good grip on that ballooned catheter and pulled it out. He didn't even hurt himself.

At the control center, the systems experts still had no answer about the capsule's prospect. There was no immediate danger. Kraft let Enos pass beyond Hawaii, then keyed a transmission to the tracking station at Point Arguello, California.

"California, Flight."

"Go, Flight." The response came from astronaut Gordon Cooper, on duty at the station.

"Get ready to bring him down. We'll let you know." Flight was shifting his mind into emergency mode. *The crew comes first. Enos is crew.*

As Enos approached the California coast, dropping toward the point where the retrofire signal would have to be sent, Flight still didn't have word from his experts.

"Systems, this is Cape Flight. What's your recommendation?"

"Hold one, Flight."

"Let's let it go, Systems. Do you agree with that?" After fifteen seconds of silence, Kraft barked it again. "Are you willing to go one more orbit?"

There was still silence. The man called Flight looked at the clock, paused a long beat, then clicked his mike again.

"You've got twelve seconds!" Everyone on the loop heard a new and harder edge to Kraft's voice. He wasn't asking anymore. He was demanding an answer. If he didn't get one, and *now,* he'd make the decision himself and some people would be looking for new jobs.

"Roger that. We oughta bring it back in!"

Kraft didn't respond. He turned his attention to California.

"All right, go ahead with retrofire on my mark." There was barely time. Kraft ran through the shortest countdown in space history. "Five, four, three, two, one, mark!"

Arnie Aldrich, the boss at Point Arguello, was ready. He sent the retro commands at precisely the right instant. Enos was on his way home after two orbits of the Earth. On that day, Christopher Kraft came of age as *the* flight director. Others would follow, but there would only be one of this kind, *the man called Flight.*

The rest was routine, except for the moments Enos vented his anger during his postflight physical in Bermuda by baring his teeth and snapping at people. It looked like he was grinning until he tried to take a chunk out of somebody's hand. Bob Gilruth and Walt Williams, slowly getting more comfortable with reporters who seemed to be learning some of the technical intricacies of covering space, explained the problems and their implications at a press conference. The thruster trouble, Williams said, wasn't serious. An astronaut could have gone to manual control and saved all that wasted fuel.

Enos wasn't trained as a pilot, so fuel depletion became a concern. All in all, Gilruth and Williams said, they were satisfied with the way things worked out.

Then Gilruth broke tradition and established a new one by actually answering the question when someone asked who would get the first orbital flight. So far, Gilruth had waited until just before launch day to make that announcement. But he'd decided that he was wrong. Mercury was an open program, not a secret one, and he understood America's need for heroes in the space race.

So he gave them two missions, not just one. John Glenn was next, backed up by Scott Carpenter. They'd try for launch on December 20, only three weeks away, though he didn't say that out loud. Getting an American in orbit before the end of 1961 would quiet the critics and give the country a smile and a cheer to carry into 1962. Deke Slayton would get the flight after that, with Wally Schirra as his backup. Reporters expected the Glenn appointment. Slayton was a happy surprise. Some of them had heard about his heart murmur. If they'd seen his smile and the light in his eyes when he got the news from Gilruth, they wouldn't have questioned the decision at all. Slayton was tough. He had heart.

The next day Enos left space insiders laughing and blushing. They brought him back to the Cape, and when he walked down the stairs from the airplane, he scared his handlers by getting that grin again that told them he was about to bite. Instead, Enos pulled loose and yanked down his diaper. For a moment, he started to fondle himself. A few reporters, photographers, and NASA officials watched in horrified fascination. But Enos had more class than that. He sat on the bottom step of the aircraft stairs, crossed his legs, leaned forward with his chin in his hand, and grinned even wider. *Gentlemen,* he seemed to be saying, *do you have any questions?*

Until he died a year later of an unrelated infection, and then lived forever in cocktail party stories and the memories of space folk, America's second chimp in space was known by his affectionate and graphic appellation.

Enos the Penis.

It only took a week to see that they weren't going to make it. Bob Gilruth told his public affairs people to put out a news release mentioning cooling system problems and saying that John Glenn would fly after the New Year. Whatever disappointment the Mercury team felt was tempered by relief that the immediate rush could slow enough to let people have at least a little time off during the holidays.

At almost the same time, NASA headquarters made the first of a series of announcements. The two-man Gemini program was approved. It was given to Bob Gilruth, who immediately established a Gemini office at his Manned Spacecraft Center, now beginning to occupy offices in Houston while its permanent base was being built. The contract to build Gemini spacecraft soon went to the McDonnell Aircraft Corporation, which was building the Mercury capsules. It wasn't long before the Air Force agreed to supply Gemini with Titan II boosters and Agena target vehicles. (The Air Force was thinking about its own Manned Orbiting Laboratory, a mini space station for both science and spying, and welcomed the chance to get the experience Gemini offered.)

About this time, there developed a semantic difference between capsules and spacecraft. The latter could maneuver to higher and lower orbits, and even sideways—*"out of plane"*—in orbit. The former could not. They followed the same looping track around the Earth, and the only maneuver they could make was retrofire to come home.

Mercury. Gemini. Apollo. America had three manned space programs on the books before Christmas 1961, and all of them were directed by a forty-eight-year-old man who looked like he was sixty and acted like he was thirty-five. Bob Gilruth was sometimes relieved that he didn't have the fame of Wernher von Braun, but still it would be nice to get some recognition. The president of the United States had his home phone number and sometimes called. But mention Bob Gilruth's name in public and the standard reaction was "Who?"

They knew who he was in the aerospace world, where all of them were discovering that the space race was no eight-to-five activity. A very few fell by the wayside and looked for other jobs. But most of them, from the astronauts who got the glory to the technicians who got none, to the wives who got ignored, to the kids who grew up while their dads aimed for the moon—these amazing people were swept up with a patriotism and a dedication never seen except in war, and as years went on, maybe not even then.

On the communist side of the world, Sergei Korolev, the Chief Designer, had no idea of the forces he'd unleashed and now gathering against him.

In another place, Bob Gilruth took a few days off, sailed in the Chesapeake, and fiddled in his living room with an idea for a swift sailing yacht with hydrofoil outriggers. When they moved to Houston next year, this pleasant and soft-spoken man told his wife, Jean, maybe they'd start building a fifty-three-footer in the garage. And one day when their special Camelot ended, they'd just get aboard, leave it behind, and sail alone around the world.

John Glenn went to the pad five times in January and February 1962, ready to ride into orbit. Three of them were complete rehearsals, going through the whole thing from getting into his space suit to getting into the cramped Mercury *Friendship 7* capsule. Once, he sat in the capsule for more than five hours, expecting to launch, but finally being told that the weather was too lousy to let him go.

That day he was the victim of a gotcha. Checking out his capsule's periscope, he saw a beautiful pinup girl's photo with the words "It's just you and me in space, John baby!" Even Glenn had to laugh. The next time he was in the capsule and looked into the periscope, the photo had been changed. Now it was an ugly scrubwoman with the caption "What did you expect after yesterday's fiasco?"

One afternoon he went for a run on the beach and saw the tracks of sea turtles coming up to nest. He mentioned it to a *Miami Herald* reporter, and the next day the reporter did a facetious column about Glenn's new recipe for turtle soup. "I got five hundred letters against me on that one," Glenn remembered with a sad shake of his head.

Other launch dates were announced, but something went wrong—from more bad weather to broken equipment—before he suited up and made that fifth trek from the spartan astronaut quarters at Cape Canaveral to the Atlas launch pad a few miles away. Along the way, Ralph Morse from *Life* noticed something strange. Each time Glenn, the "old man" and the internal conscience of the Original Seven, went to the pad, he did the same thing. Wearing his silvery space suit and carrying a portable air conditioner in his right hand, he'd pause at the door to the elevator that took him to the top of the gantry. Then he'd switch the conditioner to his left hand. Morse saw it through his camera lens and wondered.

"Why do you do that, John?" he asked after the third time.

Glenn's eyes twinkled and he was quiet for a moment. "Well, I'll tell you, Ralph," he said conspiratorially, "but you have to promise not to tell any of the *Life* writers. I don't want this in the magazine."

The high-energy photographer almost bounced off the ceiling as he vowed to keep the secret. "Okay," Glenn said, "that elevator level is the only place where there's a live TV camera that sees me up close." His voice took on a softer quality. "Annie can see me there." He paused again. "When I switch that unit to my left hand, I'm giving Annie a message. I'm telling her that I love her."

Morse swallowed a big lump in his throat and kept the secret for thirty-six years.

John Glenn told Annie he loved her at 5:25 a.m. on February 20, 1962. A few minutes later he was in *Friendship 7,* wrapped up in the details of the countdown, sitting through one hold after another until the planned launch time of 7:30 a.m. came and went, feeling his muscles cramp and wondering if this was another day that would end in frustration. It didn't. At 9:46:45 a.m., Glenn was still on the pad in Florida. Scott Carpenter in the block-house came on the loop. "Looks good, old man," he told Glenn. Four seconds later he uttered words that would be quoted around the world: "Godspeed, John Glenn. Ten, nine, eight, seven . . ."

At 9:54 a.m., Glenn was in space, five hundred miles downrange and one hundred miles up. When he reached down to open his map case, he got a dose of Al Shepard's humor. One of the Jose Jimenez bits involved a little gray mouse launched in a rocket nose cone. Now a little gray toy mouse on a slender tether, hidden there by Shepard, floated up from the map case. Glenn chuckled and tucked it away.

He pulled out his camera and took the first decent color pictures of Earth from orbit, capturing mountain ranges and forests and deserts in red and brown and green splendor. In the preflight training, the doctors had insisted that no camera be carried. They weren't yet certain that an astronaut's eyes wouldn't go all fuzzy in zero gravity—the experience of Yuri Gagarin and Gherman Titov notwithstanding. And even if he could see, they didn't want Glenn preoccupied with photography when he could be doing other things to prove or disprove their theories about man in space.

Glenn thought it over, decided the medics were being silly, and went to see Bob Gilruth. "I know people say it'll be distracting," he told Gilruth, "but heaven's sake, I'm not going to get so carried away with the camera that I'm not going to monitor the things I'm supposed to be monitoring."

Gilruth overruled the naysayers. "Talk to that *Life* guy and see what kind of camera you can get," Gilruth said.

Ralph Morse was happy to oblige. A few days later Glenn brought in a 35-mm Leica purchased by *Life* and they started practicing. It wasn't easy.

Engineers tried to rig up a handle so Glenn could use the camera one-handed, tapping the shutter button and advancing the film while wearing a space suit glove. A few days before he was finally scheduled for launch, Glenn drove into Cocoa Beach to get a haircut and to pick up a few things at a drugstore. There he spotted a new Japanese 35-mm camera, a Minolta Hi-Matic with a big shutter button and an auto-focus lens. He bought it for forty-five dollars. Now he was aiming it out the window of *Friendship 7* and taking historic photos. Later he laughed about it. "The first good pictures taken by a man in space," he grinned, "and it was a drugstore camera."

Nor did it interfere with Glenn's chores. He adjusted quickly to zero gravity, he said, and "it seemed natural to let the camera go, flip a switch, and then take the camera back out of midair." The absence of gravity, Glenn was showing to the doubters, was more friend than foe.

Back home and down below, Americans were going crazy. More than 50,000 people had lined the beach and the roads leading into Cape Canaveral to get a look at the Atlas rocket sending *Friendship 7* skyward. Another 100 million watched on television, one of the largest TV audiences of the time. Thousands crowded shoulder-to-shoulder in New York City's Grand Central Station, not wanting to walk the last few blocks to work when they could watch this drama on the large screen put there just for today.

It was dark on the other side of the world. In Perth, Australia, they turned on all the lights, everything the Aussies could find to illuminate their city. Would a man in space see them? It was a question without an answer until John Glenn looked down and told Gordo Cooper, on duty in the tracking station at Muchea, that he saw a very bright light. "That's Perth," Gordo told him. In Perth, they cheered when they heard the report.

Coming up on a new sunrise over the Pacific, Glenn saw the "fireflies," bright yellowish-green flickering specks surrounding *Friendship 7*. They disappeared in full sunlight. He'd take some kidding about them, but there was a logical answer and another astronaut would find it.

Trouble started over Mexico. Mercury's automatic control system failed. The capsule began drifting to the right, not holding its attitude. Glenn switched to manual—they called it *"fly-by-wire"*—and had no trouble bringing things back. It took some extra spurts of the hydrogen peroxide fuel, but it showed how a man could overcome these things and keep his ship upright and steady.

The old redhead felt good passing back over Cape Canaveral. Chris Kraft and his flight control team didn't share Glenn's serenity. Their instruments and incoming data from *Friendship 7* showed more trouble. Below Glenn's butt, and on the other side of the pressure shell, there was a lot of equipment. From inside to outside, there was a landing bag, programmed to inflate a few seconds before splashdown to help cushion the impact. Next came the heat shield. Finally there was the retropack of three rocket motors that would slow the capsule down and drop it from orbit. Signals showed that the landing bag was already inflated. The man called Flight asked the questions quickly and sharply.

Had it pushed the heat shield loose? If it wasn't secure, John Glenn could be cremated in the three-thousand-degree reentry fireball. Was the bag really inflated, or was the signal faulty? There was no way to know. They'd had bad switches and short circuits before, in other parts of capsules. This could be a false alarm.

Flight told his team to look at the possibility of reentering with the retropack still in place. Glenn was supposed to jettison it after the rockets fired. Could he leave it there, pressing against the heat shield, just in case? There wasn't much time. This was a three-orbit flight and Glenn was well into his second turn around the world. Flight decided not to worry the astronaut with these questions. Instead he passed the word to Gordo Cooper in Australia. "Verify that the land bag switch is off."

As Glenn came up on his second pass over Muchea, Cooper asked the question. Yes, Glenn said, the switch is off. But he was clearly puzzled. Why would they ask that kind of question? One of the controllers called Max Faget, already working out of an office in Houston. Nobody knew more about the Mercury capsule. He'd done much of its design. Could they leave the retropack in place?

"Absolutely," Faget said. "Unless . . ."

"Unless what?"

"Unless all the retros don't fire. If there's fuel left in them, it'll ignite during reentry."

The fireball would be brief, bright, and fatal.

So the word came back to Flight. The retropack could stay in place, probably, during reentry. It wasn't likely to cause big problems. But Flight wanted to check all the angles. On his last pass across the Pacific, as he approached Hawaii, Glenn heard another strange request: "Put the landing bag switch to Auto," he was told. "Do you have a light?"

"Why do they want that?" Glenn asked. Something, he knew, was wrong. So they told him. He put the landing bag switch briefly in the Auto position, and the onboard signal light stayed dark. There was a brief sigh of relief in the control center at the Cape. It meant that the signal itself was probably bad; the landing bag probably had not inflated.

The key word was *probably*.

Now the man called Flight had to make a series of decisions. In the few minutes left as Glenn crossed between Hawaii and California, what Flight decided could mean life or death for John Glenn. Flight was ready for the responsibility. *The crew comes first. Glenn is crew.*

Shorty Powers was keeping the world informed about the tense situation. Things were not A-OK with John Glenn. The American TV audience grew far beyond 100 million. Around the world, hundreds of millions more drew up to their radios to follow this real-time American drama. Flight keyed his mike and made his first decision. He told Wally Schirra at the Point Arguello tracking site in California to tell Glenn to leave the retropack in place through the first part of reentry. Schirra passed the word: "Keep the retropack until you're over Texas," he radioed.

"Okay," Glenn acknowledged. He was too busy to ask why. He set up *Friendship 7* for retrofire, using his fly-by-wire control to position it perfectly. Schirra echoed the countdown to make certain that Glenn heard it.

". . . three, two, one, retrofire!"

Number one fired. Number two fired. Number three fired. It was perfect. Eliminate one of the *probably*s.

"All three fired," Glenn radioed. "It feels like I'm going back toward Hawaii."

"Don't do that, John," Schirra chuckled. "We need you on the East Coast."

They had him on radar and could see *Friendship 7*'s trajectory falling perfectly on the ordained line. In the control center, Flight made one more decision.

"Leave the retropack on through the entire reentry," Schirra radioed.

"Why do they want that?" Glenn queried.

"This is the judgment of Cape Flight."

That was good enough for John Glenn. He'd seen Chris Kraft at work, handling simulated emergencies from an astronaut having a heart attack to the world going to war with atomic bombs during a mission, and real emergencies on the monkey flights. If the man called Flight wanted him to keep the retropack in place, that was what he'd do. Moments later, the three-thousand-degree heat of reentry put an envelope of ionized air around the capsule. No radio signals could get in or out for the next several minutes. The world stopped to hear John Glenn's fate. If he was alive, he'd crossed the Gulf of Mexico, then Florida, and was now dropping almost straight down over the Atlantic.

Al Shepard was in the control center. *"Friendship 7,* this is Cape. Do you read?"

Silence.

"Friendship 7, do you read?"

Silence.

Shepard's voice was getting agitated. *"Friendship 7,* this is Cape, do you read?"

"Loud and clear, how me?"

"Reading you loud and clear!" Shepard almost shouted. "How you doing?"

Now there came a weariness in Glenn's voice, a low note that hadn't been there before. "Oh . . . pretty good." *No,* the undertone said, *not so good.* In fact, it had been a helluva ride down from space. Glenn knew he sounded troubled, so he picked up the tempo. "My condition is good, but that was a real fireball. I had great chunks of that retropack breaking off all through." He didn't say that he'd run out of control fuel and that the capsule was jerking and swinging left-right far out of acceptable conditions. During debriefings, he compared it to a "falling leaf," a swaying maneuver pilots use to lose altitude rapidly. It is not a comfortable maneuver.

The sound and the fire had been fearsome. But *Friendship 7* held together. The heat shield did its job. Now the parachutes were coming out and Glenn's enthusiasm returned full force. Even the landing bag came out on schedule. He splashed down forty miles short of the recovery forces. Still, it only took seventeen minutes for a helicopter to hook up to the capsule, lift it from the sea, and set it gently on the deck of the destroyer *Noa.* By then, John Glenn was hot, thirsty, exhausted, and safe.

America didn't just cheer. It howled its exultation. John Glenn was a hero and America was closing the space race gap with those Russians. Jack Kennedy honored Glenn in a Rose Garden ceremony at the White House. LBJ sat in a convertible with John and Annie Glenn, followed by the other astronauts and the nameless men of Mercury, to accept the cheers of a quarter million people lining Pennsylvania Avenue. Even Bob Gilruth got some attention, hailed in the press for being a man who held fast and got the job done. In New York City, the mayor declared a John Glenn Day and gave medals to both Gilruth and Glenn.

When they tore apart *Friendship 7,* they found a loose switch. The landing bag had been just fine all along. Then the United States did something that the Russians would never think of doing. NASA sent *Friendship 7* on a tour around the world. It stopped in seventeen countries. They lost count of how many millions of people stood in line, in good weather and bad, to look at John Glenn's capsule, to marvel at the space age in which they all were citizens. The world saw the difference close-up between an open program like Mercury where the emergencies and the successes played out in real time for all to see and hear, and a secret program like the Russians where they only revealed what they wanted to reveal.

When *Friendship 7* came home, it stopped for a while at the 1962 Seattle World's Fair, where the most prominent structure was a new tower called the Space Needle and where a million more people lined up to see John Glenn's ship. A young NASA public affairs guy was there to hand out literature and to answer questions. He kept looking over his shoulder at *Friendship 7,* then in awe at the never-ending line of shiny-eyed people streaming into the exhibit hall.

"Who'da thunk it?" Bob Button said quietly and more than once. "We're gonna win this race."

Deke Slayton was up next. He never forgave Jerry Wiesner for what happened.

In Slayton's version, the White House science adviser was a vindictive yellow-belly who had lost out every time he argued against any kind of a manned space program. He saw his chance to get even after the excitement of John Glenn's flight. Slayton was his fall guy. Wiesner went to Jim Webb and insisted that the astronaut be pulled from the upcoming Mercury-Atlas 7 flight because of his flaky heartbeat. No matter that the flight surgeons had cleared Slayton, or that he routinely flew high-performance jets, or that he was supremely qualified to be an astronaut.

A tragedy with Slayton would rub off on all of them, but mostly on the president, Wiesner argued. The nation can't afford to take the chance. It was one of the few times that Webb caved in to political pressure. The official record doesn't say that, but the chain of events showed that he fought almost all the way on Slayton's behalf. First, he asked Bob Gilruth for a medical review of the astronaut's history. The Mercury flight surgeons did, and cleared him once more. To nail it down even tighter, two of the top Air Force physicians at the Pentagon were briefed. They endorsed the decision. Slayton should fly the next orbital mission.

Somehow the Air Force surgeon general got into the act. Did Wiesner apply pressure? No one said so. But the surgeon general ordered a new review board, naming eight senior flight surgeons to study the file. Slayton himself appeared before the board, as did Dr. Charles Berry, the chief Manned Spacecraft Center physician. Again Slayton got a green light. He could fly.

Then inexplicably, it seemed that Jim Webb caved in. Three groups of experts siding with Slayton were not enough. He ordered a new board, not of flight surgeons, but of three civilian heart specialists. And finally Slayton was out.

The cardiologists didn't say that Slayton's heart flutter—now occurring once every two weeks or so—would interfere with his ability to work in space. What they said was that it would be better to send up an astronaut who didn't have that condition. Webb ordered it to happen. Deke Slayton's backup man, Wally Schirra, should have taken his place. But that didn't work out either.

When Chris Kraft heard that Scott Carpenter would get the flight, he was appalled. Mercury operations director Walt Williams argued that, since the flight was a near-duplicate of MA-6, it should be assigned to John Glenn's backup. Bob Gilruth agreed.

The man called Flight remembered Carpenter's abysmal performance in the control center before Al Shepard's flight and hadn't been impressed with what he'd seen since. He told Williams bluntly that Carpenter was a bad choice, that he shouldn't be allowed to fly on any mission. Chris Kraft lost that argument. It didn't make him feel any better when Scott Carpenter's flight convinced him that he was right.

Gherman Titov came to town, the first cosmonaut to visit the United States. He didn't have anything good to say about the United States or its space program. Whatever he was shown, he countered in broken but understandable English with a comment that something similar existed in the Soviet Union, but was bigger and better. When a reporter asked if cosmonauts would like to fly in American spacecraft, his laugh was harsh. "No," he said. "Quality not good enough."

Only one American sight left Titov with nothing to say. His NASA guides took him to the Library of Congress to see the Constitution of the United States of America. He studied it carefully and listened to a translation of the short lecture that everyone who passed here was privileged to hear. Then he nodded and looked up and with the barest trace of a smile slowly walked out to the car.

Wernher von Braun appeared at a reception for Titov. Maybe it was the cocktails, but he answered a reporter's question honestly. A few in the press harbored the hope that the Russian spaceflights were phony and the reporter asked how von Braun was so sure that Titov had been in space. "We watched him on television," von Braun said, confiding that NASA monitored the Russian radio frequencies and was routinely decoding their TV transmissions. The reporter was delighted with his scoop. Von Braun had forgotten that the American eavesdropping was still classified as a military secret.

John Glenn was Titov's astronaut host in Washington, and in private the two became friends. It was always that way. Whenever astronauts and cosmonauts came together—at the Paris Air Show, on goodwill tours, at scientific conferences somewhere in the world—they'd get together in private, toast each other with too many vodkas or Scotches, and trade flying stories. They were pilots, and all pilots share the bonds of flight. And they were spacemen too, so the bonds were even tighter.

"I hear this thing barbecue," Titov said to Glenn. "It is American thing, yes?"

"Yes," answered John Glenn, and explained the American custom of outdoor cooking.

"Good," Titov said. "I come tonight to your house. You do barbecue."

Glenn glanced at his watch. It was almost 4 p.m. There was just time enough. "Okay," he said. "My house tonight."

While Glenn rushed home to get ready, sending Annie shopping for steaks and hamburger, calling two neighbors to borrow their grills so they'd have three barbecues going at once, Paul Haney of the public affairs office tried to corral the Russians. Cosmonauts didn't travel alone. Gherman Titov had an entourage, including KGB men and a three-star Russian general who the NASA people thought might be the equivalent of Bob Gilruth. The

Soviet ambassador, Anatoly Dobrynin, had to be notified, and when he was, he invited himself to the Glenn house too.

Glenn lived in suburban Virginia. Haney took the point and led a caravan of high-ranking Soviets and some NASA people through the curving roads to Glenn's house. "It was the scenic route," Haney remembered. "We didn't want to get there before John was ready for us."

Finally he couldn't delay anymore and Haney turned onto Glenn's street. There was smoke coming from the astronaut's garage. "He had three barbecues going and one of the neighbors had just poured gasoline on a grill. The paint on John's garage was starting to burn," Haney said. "Everybody piled out and grabbed hoses, whatever they could find. There was no language barrier here."

The fire was out quickly. Gherman Titov walked over and put his arm around John Glenn's shoulder.

"Tell me," he asked slyly, "every time you have cookout, you burn down house?"

The house that might have burned was NASA's. Chris Kraft thought it was Scott Carpenter's fault. The astronaut had a very different view.

A few weeks after Glenn's barbecue, Carpenter was in space on a three-orbit mission. Of the Original Seven, Carpenter was the poet and the dreamer. He wanted to make the flight of *Aurora 7* more than an engineering exercise. He wanted to do some science and he wanted to feel the experience of flying in space. In Kraft's analysis, that attitude almost killed him.

In his first orbit, Carpenter didn't notice a discrepancy between where his gyroscopic instrument said his heading was and where it really was. That would cause trouble four hours later, and from his vantage point in the control center, Kraft would conclude that the real problem was with Carpenter's concentration. "It was an instrument with two needles, one for pitch and roll, and one for yaw," Carpenter recalled. "You can't see a yaw error by a quick glance out the window, and you may never see it if it's intermittent. Maybe there should have been a yaw check in the flight plan, but there wasn't."

Then he had trouble with his camera, a new 35-mm with a motorized film advance, and got behind on other assignments. When he got the camera working, he blipped his thrusters this way and that, turning his *Aurora 7* capsule to get better views of the Earth below and stars above. Part of it was approved; Carpenter's flight plan called for him to fully test Mercury's attitude thrusters. But part was his own joy at the marvelous views of Earth and at how easy it was to make *Aurora* point in any direction he wished.

Kraft was now in his fourth manned mission as Flight. He was aghast when *Aurora 7* passed back over Cape Canaveral and his control team read out the capsule's condition. Carpenter's fuel already was getting low. Even a simple maneuver like pointing the nose backward instead of forward could use two pounds of the precious hydrogen peroxide thruster fuel. Flight's order went out to the tracking network: tell him to stop using fuel. But getting the message to Carpenter wasn't that easy. The network still wasn't tied together with real-time communications on every channel. Some messages went by Teletype. There was a Capcom at the Cape, but he could only talk to Carpenter during passes over the United States and the near-Atlantic. At key sites around the world, other astronauts served as Capcoms and relayed messages to the orbiting capsule.

Either Carpenter didn't get a clear order to cool it, or he ignored it. That would be another point of contention. And there was another cooling problem at the same time. Temperature readings in mission control showed the capsule, and Carpenter, getting too warm. Carpenter wasn't sure whether it was the suit cooler or the cabin cooler causing the problem. Checking them took time and didn't seem to make much difference. But it was putting him more and more behind schedule. Then while he was tending to his workload —star sightings, trying out food samples, looking for flares fired into the air in Australia—he accidentally turned on a redundant manual control system. It doubled his fuel use until he saw his error. It was easy to do; the control stick was poorly designed for zero gravity, and even a slight extra movement of the hand could cause trouble. They changed the design for the remainder of the Mercury program.

In mission control, they saw that Carpenter's fuel was dangerously low as he completed his second orbit. They didn't know about the yaw needle failure. It wasn't hooked up to the radiotelemetry system. This time Carpenter got his order directly from Gus Grissom, the Cape Capcom, and it was unmistakable: use no fuel at all until it was time to set up for retrofire. In effect, Carpenter was told to let *Aurora 7*'s attitude drift during his third orbit. That solved the fuel problem, at least for a while, but led to a new one.

More than halfway around the world, Carpenter stowed his camera on the hatch, and as it banged home, thousands of John Glenn's fireflies magically appeared outside. He banged it again and there were more. His third sunrise was behind his back and the lighting was perfect to see what was happening. The fireflies weren't some kind of life in space; a coating of frost on the capsule's outside shingles, gathered there each time moisture-laden air was vented, was the culprit. When Carpenter banged the bulkhead, they flew off and glittered, but only at sunrise when the angle of incoming light was just right.

It was a beautiful sight. It was too bad that Carpenter solved the firefly mystery at the same time he was supposed to be getting ready for retrofire. Suddenly he was within range of the Hawaii tracking site and he hadn't gotten far along on the countdown checklist. When he tried to hurry through it, he missed a few things. As he passed Hawaii and headed for California, trackers told mission control at the Cape that he seemed confused or preoccupied. "The fireflies slowed me down on my retrofire preparations," Carpenter remembered. "I *was* preoccupied with the discovery, but I wasn't confused."

Nobody yet knew that in addition to the faulty yaw needle, the critical gyro horizon instrument was now misaligned. When Carpenter discovered the problem, he realized that the automatic control system wouldn't work for retrofire. At Point Arguello, controllers in the tracking station could see on their instruments that something wasn't right. It was up to Al Shepard, in the California Capcom seat, to alert Carpenter.

"Seven, this is Capcom. Are you in retro attitude?"

"Yes. I don't have agreement with ASCS in the window, Al."

Translation: Carpenter could see that the outside horizon's position, seen against etch marks on the window, didn't match up with his automatic stabilization and control system readings. "I think I'm going to have to go to fly-by-wire and use the window and the scope. ASCS is bad. I'm on fly-by-wire and manual." He was trying to set up *Aurora*'s position manually by looking through the window and the periscope.

"Roger, we concur," Shepard answered. "About thirty seconds to go" (to arm the retrorockets). At the same time, ground controllers saw that Carpenter hadn't put switches in the right positions for a manual retrofire. Shepard counted down to aid Carpenter in arming the retros. Then he moved on the other problem. "Check ASCS quickly to see if the orientation mode will hold . . . if your gyros are off, you'll have to use attitude bypass."

"Gyros are off," Carpenter replied. He didn't catch the message about checking the automatic system, and the moment of retrofire was approaching rapidly. "I should have thought about it myself, but I didn't," he said later.

"You'll have to use attitude bypass and manual override," Shepard called urgently. Carpenter was trying to align *Aurora* manually and there wasn't time to get it precisely right. Everything was coming up fast. Shepard gave him the retrofire countdown, and by the time Carpenter got the switch setting changed, the retrorockets fired three seconds after Shepard called out "zero!"

"Okay, fire one, fire two, and fire three," Carpenter radioed. "I had to punch off manually. I have a little bit of smoke in the capsule."

"Attitude hold, Scotty," Shepard warned. It was critical over the next several minutes that Carpenter keep the capsule precisely aligned for the plunge into the atmosphere.

"Okay, I think they held well, Al . . . I think they were good."

They weren't and it was about to get worse. Scott Carpenter was going to miss the selected landing spot in the Atlantic. He was on a trajectory to land long, by a lot. Those three seconds of late retrofire helped; the attitude misalignment did the rest. Any landing assumed he survived the coming fireball. For that, he either needed fuel enough to use his control thrusters or a lot of good luck. From the conversation between Carpenter and Shepard, it sounded like he didn't know the difference.

"How are you doing in reentry attitude?" Shepard asked.

"Stowing a few things first," Carpenter answered.

A half minute later Carpenter came back on the air: "Going to be tight on fuel."

"Roger, you have plenty of time," Shepard answered.

Carpenter's next transmission came from the poet, not the test pilot. "Okay, I can make out very, very small . . . farmland, pastureland below," he said casually. "I see individual fields, rivers, lakes, roads, I think. I'll get back in reentry attitude."

Shepard was getting frustrated. "Roger, Seven," Shepard called, keeping his voice calm, "recommend you get along to reentry attitude, using as little fuel as possible . . ." Thirty seconds later Carpenter was out of radio range. It was up to Gus Grissom at the Cape to handle it from there. Flight told Grissom to start with the basics. No matter that Kraft would vent his anger later. Now was the time for icy calm and careful thought. *The crew comes first. Carpenter is crew.*

Grissom did his job. "Do you have your faceplate closed?" he asked Carpenter.

"Negative. It is now. Thank you."

The fireball was coming and Carpenter was just finishing his checklist. Then came the fire. He was quickly out of communication and on his own. For several minutes, he talked to the onboard tape recorder. "I hope we have enough fuel . . . I get the orange glow at this time . . . bright orange glow . . . my fuel, I hope, holds out . . ."

It didn't. It was all but gone. Outside, the fireball surrounded *Aurora 7,* and pieces of heat shield were flying off. Carpenter was amazingly calm, proving his bravery and even cracking a Jose Jimenez joke. "There was a large flame piece coming off," he said to the tape recorder. "Almost looking like it came off the tower." Then mimicking the Jose Jimenez line when asked if this piece of headgear was called a crash helmet, Carpenter recorded the immortal line "Oh, I hope not."

Now *Aurora 7* was beginning to swing back and forth, going even beyond the oscillations that John Glenn experienced. Fuel was gone. Carpenter had no way to control his capsule, nothing to do but hang on and wait it out. He smoked into the denser atmosphere and the fireball disappeared. But the violent oscillations continued. "Think I'd better take a try on the drogue," he said to the recorder. By manually deploying the drogue parachute, he decided, the oscillations would stop. "Drogue out manually at twenty-five [thousand feet]. It's holding and it was just in time."

The main chute followed, and for a few seconds, Grissom got him on the radio. "I'm on the main chute at five thousand, status is good," Carpenter reported. "I am not in contact with any recovery forces. Do you have any information on the recovery time?" As the capsule dropped toward the sea, radio static increased. There was time for one quick advisory.

"*Aurora 7, Aurora 7,* Cape Capcom, be advised your landing is long. We will jump air rescue people to you in about one hour."

"Roger, understand, one hour."

Then the radios went to solid static. Neither Carpenter nor Grissom could hear the other.

Now the world waited in another fit of agony. No one outside of the control center and the tracking stations knew why Carpenter had missed the prime recovery area. They only knew that he'd overshot, was out there somewhere, and that an air rescue search was under way. Again the American TV audience approached 100 million, all of them frozen in suspense. It was a classic case of civilians not understanding how the military and NASA worked. Carpenter wasn't lost. He'd landed 250 miles long and it would take a while for an airplane to get there. He was spotted thirty-six minutes later, laid-back in his life raft, drinking water and eating a candy bar. The message went out: The astronaut was safe.

Carpenter saw the planes, but not the first rescue jumpers going out behind him. When a frogman suddenly appeared at the edge of his raft, Carpenter gave a start: "How did *you* get here?" he asked in surprise. He was glad to see him. Getting the recovery equipment and helicopters to the spot took hours. When Carpenter was finally deposited on the aircraft carrier *Intrepid*, he made headlines around the world. The frogmen had told him that his location had been initially unknown. That was wrong, but when a news pool reporter on the carrier asked about it, the astronaut repeated the story.

"I didn't know where I was, and they didn't either," Carpenter said. The remark gave NASA a temporary black eye, but agency people quickly convinced other reporters that Carpenter's location was never in doubt. They'd tracked him all the way to the horizon. When the reporters heard Grissom's last transmission, they knew it was the truth.

But the man called Flight was livid. In his eyes, Carpenter had been capricious and careless during this three-orbit flight, and then mouthed off to a reporter. Chris Kraft swore an oath that day and made it stick: "Scott Carpenter will never get another flight as long as I have anything to say about it."

His oath was private, at least for a few years, and the American public rightfully saw Carpenter as another astronaut hero. Rene Carpenter admired how he handled both the internal strife and the external adulation. Sometime after the flight of *Aurora 7*, she talked to a Rotary Club luncheon in Austin, Texas. All of these astronauts, she said, had a "maturity that helped them handle the fantastic public acclaim and that great American game called Here Comes the Hero." Then she put a military wife's perspective on things. "Don't ask the old hackneyed questions, 'Aren't you afraid?' and 'How can you stand to let him do it?'

"We do no more than the wives of helicopter crews in Vietnam, or the women the *Thresher* left behind. Don't feel sorry for us—for the tension and the waiting. The rewards are great—great to be able to whisper at liftoff 'Don't look back. We're with you.' "

She was with Scott Carpenter when Kraft made his oath stick, and there were no more space missions in the astronaut's future. While he was still waiting and hoping, Carpenter focused on underwater training to simulate weightlessness during space walks and on related tasks. Just before he went back to the Navy to work on, then lead, the underwater Sealab projects, he helped install a huge training water tank at the Houston space center.

Kraft and Carpenter would never agree on the truth of *Aurora 7*. But Carpenter left gracefully and felt that he'd left something valuable behind.

Water cooler talk said that an unidentified Air Force general was coming to run the astronaut office. It wouldn't be long before a second group of astronauts was brought into the program, to help fly Gemini and Apollo missions, so it was logical that NASA wanted some muscle to keep them in line.

There was plenty of muscle already in place. Deke Slayton had showed both loyalty and class in how he'd handled getting dumped from his space-flight assignment. After some lobbying by the other astronauts, and agreement from key people like Walt Williams, Max Faget, and Chris Kraft, he got the job.

When he named Slayton as chief of the astronaut office, Bob Gilruth knew he was getting one tough son of a bitch. He didn't know that Slayton would evolve into a force that sometimes equaled his own, or that Slayton's growing power would one day make him the man—virtually without threat of veto—who decided who flew, who didn't, and when and why. The astronauts loved it. Slayton made the decisions and Slayton took the heat. Four words were often enough to fend off an unwanted request:

"Hey," the astronaut being queried would shrug, "go see Deke."

The baby steps of Mercury got all the public attention. With every mission, something new and strange happened, and Walt Williams at the Cape marveled at how much they still had to learn. Away from the reporters, the televised launches, and the nail-biting flights, fierce arguments were raging about how, exactly, one goes to the moon. The answers couldn't be found by trial and error, and the choices made in 1961 and '62 had to be converted into hardware, mission plans, and training for ground crews and astronauts alike.

It didn't take long to eliminate direct ascent from the lunar schemes. Within weeks of Kennedy's moon speech, a NASA technical team estimated that a three-man Apollo ship that could land on the moon and get back home again would weigh seventy-five tons. Mercury weighed less than two tons, and it taxed the limits of an Atlas rocket just to get it into orbit around Earth. Even the apocryphal Nova rocket, with more than eleven million pounds of thrust, wouldn't be enough to send Apollo straight from Cape Canaveral to some hopefully level lunar *"mare,"* or sea.

The obvious answer was being pushed by Wernher von Braun and his team at the Marshall Space Flight Center: Earth orbit rendezvous. Von Braun's growing empire would provide the rockets to put Bob Gilruth's Apollo pieces into orbit where they could be assembled, then sent on to the moon. Each mission would take two to four rocket launches, depending on the Apollo configuration. And if von Braun could undercut Gilruth by getting part of the spacecraft assigned to Marshall, so much the better.

Gilruth's people weren't so sure. Max Faget was having trouble getting a design for an Apollo spacecraft that went directly from Earth orbit to a lunar landing. It was a huge ship, and no matter whether it landed on skids like an airplane or backed down on braking rockets like a Chesley Bonestell painting, the technology challenges were enormous. Other ideas were being tossed about, but none seemed to make it any easier. One fellow argued otherwise. John Houbolt had the innocuous title of associate chief of the Dynamics Loads Division at Langley Research Center. He was far down the chain of command and without much influence. That didn't stop him. He'd been studying rendezvous in space, and with some of the people working for him, John Houbolt came up with something new.

He called it lunar orbit rendezvous (LOR), and everyone who heard about it thought he was nuts. In the LOR scheme, it only took one large rocket from Wernher von Braun's inventory. The Saturn 5, agreed to by von Braun and Gilruth in late 1961 and approved by NASA headquarters, could do the job. The Saturn 5 clustered five of von Braun's F-1 rocket engines in its first stage to generate 7.5 million pounds of thrust. With smaller second and third stages, Houbolt thought that was enough to make a manned moon mission possible.

Houbolt's vision saw an Apollo spacecraft with a main, or command, module (CM) carrying three astronauts, and a second unit, the lunar excursion module (LEM), that would land two of them on the moon. The command module would stay in lunar orbit. When it was time to go home, the top half of the LEM would blast up with its own small rocket engine and rendezvous *in lunar orbit* with the command module. The astronauts would abandon the LEM stage and blast for home with the CM's rocket.

It was the rendezvous in lunar orbit that scared people. The United States still hadn't put John Glenn into Earth orbit. Rendezvous involved tricky maneuvers that were still theory, had never been tried and wouldn't be for years. Now this guy Houbolt was trying to convince the bosses that lunar orbit rendezvous was the way to go, but almost everyone believed the plan was a recipe for dead astronauts. Houbolt refused to forget it. He stepped out of channels in May 1961 to send a letter to Dr. Robert Seamans, NASA's associate administrator and number three man, urging him to look carefully at all rendezvous schemes. Seamans didn't pay much attention until November, when Houbolt got really out of line by sending him a nine-page letter that challenged all of the rendezvous thinking current in NASA, including Wernher von Braun's. Houbolt outlined his attempts to get serious people to consider lunar orbit rendezvous and asked Seamans a blunt question: "Do we want to get to the moon or not?" He underlined the question.

Then he complained about an unnamed senior headquarters person who commented that "Houbolt has a scheme that has a 50 percent chance of getting a man to the moon, and a 1 percent chance of getting him back." That comment, Houbolt wrote, exhibited "stupidity." Then he stopped beating around the bush and got blunt with Seamans.

One of the analytical methods used in rendezvous evaluation involved "self-hypnosis," he wrote disparagingly. Some booster development meetings were shams, he said, because people refused to believe the design numbers, or the meetings fell apart when somebody just shook his head and walked out. He ragged and nagged. But he'd gotten Seamans' attention.

Finally on page nine of his letter, John Houbolt got to the point. They needed to try out rendezvous in the Mercury Mark II program (soon to become Gemini), focus rocket engine development on a heavyweight booster (soon to become the Saturn 5), and firmly commit NASA to lunar rendezvous. "Give us the go-ahead and we will put men on the moon in very short order," he wrote, "and we don't need any Houston empire to do it."

On that last, he was wrong. NASA did need the Houston empire taking shape as the Manned Spacecraft Center. Still, Seamans was impressed enough with Houbolt's audacity to begin pressing for more study on lunar rendezvous. Neither knew that the idea was already taking root in Bob Gilruth's staff, so Seamans' support from the Washington headquarters just gave it an added push. Max Faget had listened to an LOR presentation, went away scoffing, then sat down in his office and reran the numbers. The second time around, LOR looked pretty good. A rendezvous was a rendezvous, and if they perfected the technique in orbit around the Earth during the Gemini program, there was no reason for it to be more dangerous around the moon.

Faget saw that LOR solved the problem of how to land on the moon, or at least made it less daunting. Instead of landing a huge ship there, the lunar excursion module would be smaller and more manageable. Backing it all up from the moon to the launch pad, Faget saw that a Saturn 5 booster just might be big enough to pull off a moon mission with a single rocket launch. LOR wasn't complicated, Faget realized. It made a lunar landing simpler and cheaper too. He tried out his new thinking on Bob Gilruth and the boss agreed. "I'd talked to Houbolt myself, and from an engineering view, he seemed to make sense," Gilruth said later. From that moment in early 1962, the Manned Spacecraft Center took the lead in pushing for LOR.

None of that sat well with Wernher von Braun. The Saturn 5 was an interim rocket in his eyes. The real moon booster would be the Nova, and it would be bigger than any rocket ever imagined. Bob Gilruth set out to change his mind. He had an elaborate presentation put together by Charles Frick of the Apollo team in Houston, and sent it out to evangelize. The "Charlie Frick Road Show" briefed von Braun and his staff in April, and Frick reported that Wernher's antagonism was more muted when the briefing ended.

The next month, with John Glenn's mission behind and Scott Carpenter's just ahead, Frick briefed the new head of manned spaceflight at headquarters, an RCA import named D. Brainerd Holmes. Just after Carpenter flew, he did it again for Seamans. LOR was collecting converts with every presentation. One of them was John Paup, the Apollo project manager at North American Aviation, which had gotten the contract to build the Apollo command module, was hoping to be the prime contractor for the entire ship, and was backing von Braun's Earth rendezvous method. If there was going to be that separate lunar excursion module, Paup saw, North American probably wouldn't be allowed to build it. Some other company would get a fat contract. It didn't matter. He believed what he believed, and after his team heard the Frick briefing, he stood.

"We are backing the wrong thing," he said. "Who is the son of a bitch that is against LOR here?" It wasn't good for North American, but every engineer in the room knew that it was the right thing to do.

At the same time, another headquarters executive, Joe Shea, had asked the rocket experts at Marshall Space Flight Center for a detailed comparison between the Earth and lunar rendezvous schemes. It all came to a head on June 7, 1962, in Huntsville. All day long, through sometimes rancorous questioning and discussions, von Braun's people made the case for EOR. Max Faget was there, along with a handful of Houston engineers, and they made their own case for LOR, usually by poking holes in the Marshall data. Wernher von Braun sat and said nothing as the two sides argued. When his people finished, and were satisfied that they'd made the case for EOR and that Joe Shea would carry that recommendation back to Washington, von Braun stood.

He'd listened, he'd weighed the evidence, and now his people expected him to make the closing argument that would win the day. But von Braun was a rocket scientist first, a bureaucrat second. "It is the position of Marshall Space Flight Center," von Braun began, and everyone in the room leaned forward, "that we support the lunar orbit rendezvous plan . . ."

There was stunned silence as he paused. Had they heard right? Without so much as a minute's consultation with his staff, was Wernher von Braun switching sides?

He was. United States astronauts were going to the moon, and they'd do it by following John Houbolt's crazy scheme. Lunar orbit rendezvous was now official NASA policy.

All but a handful of the Space Task Group's eleven hundred people made the move from Langley Research Center, Virginia, to Houston in 1962. Now they were called the Manned Spacecraft Center, but it was only a name. The physical site of the new center had, as Bob Gilruth promised, dried out. Surveyors and construction crews were busy platting the ground and digging the first holes for its buildings.

Gilruth hadn't been entirely happy with the move. At Langley, he'd had quick access to Chesapeake Bay, where he and Jean sailed, often trying out his newest design for a hydrofoil hull. The yachting crowd knew him well after a quarter century on the water. He couldn't get anybody to race him, except the occasional new guy who had to see for himself what the old man's boats could do. What they could do was go faster than anything under sail on the Chesapeake. If Bob Gilruth wasn't so enamored of airplanes and spacecraft, he might have made a career as a boat designer and ended up contesting the America's Cup. Instead he made it a hobby and a release valve from the pressures of his real job.

When he bought the charts for the Galveston Bay area, he made a discovery. No more than eight miles south of the Manned Spacecraft Center's land, Dickinson Bayou snaked through the flat terrain. It was a broad stream and deep enough for the boats Bob Gilruth designed. More important, the bayou emptied into Galveston Bay close to Texas City. And Galveston Bay led directly to the Gulf of Mexico.

He had to see for himself. The charts could be wrong. On a trip to Texas, Gilruth rented a boat and ran up Dickinson Bayou on a personal exploration. The bayou was as charted, but the land was ugly. The first five miles west from Galveston Bay weaved through marsh and scrub. He saw bleak muskrat, snowy egrets, fat little birds with yellow beaks that he didn't recognize, even a middlin'-size alligator sunning on a mud bank. Where the grass was shorter along the banks, he could see cattle grazing in sunbaked fields. Up ahead were trees.

The water wasn't deep, but he pressed on to look for the kind of place he had in mind. It was shady under the trees, mostly oak and native pecan and a few pine. This was more what he was hoping for. And around a bend, he found it. He could see a road through the trees. For the next hour, he walked this land, liked it, and made certain he could find it again from the road.

Max Faget went along and they surveyed to be sure that the image in Gilruth's mind translated to reality on the ground. When he was convinced that it would work, Gilruth made an offer and bought the land. Faget bought another chunk a mile or so away, also on the bayou. He didn't need what his friend and boss needed; he just liked the trees and the water flowing by his backyard property line.

What Bob Gilruth did left the locals wondering about what kind of people these spacemen were. He designed a house and a garage, situating it just so on the property. The garage was huge, with doors at both ends. It was more like a warehouse or a large shop than a garage, but he could do what he wanted, it was tucked away in trees, and if people wondered, they were Texan enough not to complain.

It was a shop. Over the next twelve years, a fifty-two-foot oceangoing yacht would take shape there. It was supposed to be fifty-three feet, but the contractor goofed and made the garage a foot shorter than the plans. It was easier to change the boat than to rebuild the garage, Gilruth liked to joke. Hand-built during evenings and weekends, with only a little help from Jean, this yacht was Bob Gilruth's dream, his passion, and his outlet from the growing pressures of sending men to the moon.

First he built a one-seventh-scale model of the yacht, with its hydrofoil hull and big hydrofoil outriggers, but never bothered to draw up a set of plans. They were in his head. He sailed the model on Dickinson Bayou, measuring its responses to the wind, timing it over precise distances, using the data to modify here and tweak there until he had exactly what he wanted. The full-size yacht would do fifteen knots under sail, but draw only eighteen inches of water because of its unique hydrofoil outriggers.

Then he started to build, laying the hull, forming the outriggers. As the years went by, the shop filled with this boat. Visitors could find Gilruth hunkered down in its cabin, happily using his hands and his mind to create his other dream. By day, he focused on the moon. By night, he was an old-fashioned shipbuilder.

"The boat won't tip over," he told a young reporter. "It's designed to right itself if it goes over on a side."

The interior would be luxurious, sleeping eight and having amenities ranging from a fireplace to a special hatch for the Gilruths' pet raccoon. The jars of pickled copperheads and coral snakes on garage shelves, caught in the backyard and carefully preserved, would be left behind. It was a sailboat, with big A-frame masts, but it would also have a 120-horsepower inboard-outboard engine; the outriggers doubled as fuel tanks.

"It's being built as a long-range ocean-cruising boat," he said. That around-the-world trip after retirement was no idle dream.

Now and then Gilruth would walk the few hundred feet from the back door of the shop to the bayou's edge. He'd look at the oaks, sometimes measure their growth. But he'd taken all that into consideration in 1962. From the shop to the bayou, it looked like a forest. But there was a way through. He had planned it carefully.

When the day came to launch his yacht, he'd open those big doors and roll the boat through the forest. It would turn here, bend there, and it would reach the water without the need to cut down a single tree.

If I can lead men to the moon, he thought, *I can sure as heck lead my boat to water.*

The Chief Designer had exhausted his spacecraft inventory with Gherman Titov's flight in August 1961. It would take nearly a year to build a new set of Vostoks that incorporated changes allowing the capsule a longer lifetime in space and to give the cosmonaut Mercury-like control over where his ship was pointed. John Glenn and Scott Carpenter both flew in space while the Russians did nothing. They were modestly active in launching new Kosmos-class satellites, mostly carrying scientific instruments, but some of those failed to reach orbit and the launch record was not good. The industrial might of the United States was beginning to dominate.

In 1961, the United States tried for space forty-one times and succeeded on twenty-nine, including the suborbital flights by Al Shepard and Gus Grissom. The Soviet Union succeeded on five launches, failing four times, for a total of only nine. American satellites in orbit included the first true communications birds, Navy navigation satellites, a pretty good weather satellite, a satellite designed specifically for amateur radio operators, and a covey of early spy satellites. Film from some of those was deorbited and snagged in midair by C-130 aircraft flying the waters off Hawaii. The images were as good as anything Francis Gary Powers brought home after violating Soviet airspace in a U-2 aircraft. Four of the Soviet successes came on two dog missions and two manned missions. The fifth was a partial success, an attempt to reach Venus. Venera 1 left Earth orbit, but its signal was lost at 4.7 million miles.

Did the numbers matter to the American public, press, or politicians? Not a bit. When an American rocket exploded, it was usually front-page news. When a satellite reached orbit, the story was buried on inside pages. The space race wasn't about satellites, or about developing technologies to exploit the unique vantage point of being out there and looking back at the home planet.

The space race was about men. It was about putting men into capsules and sending them into orbit. It was about winning the moon and planting a flag and leaving footprints in lunar dust. So when Sergei Korolev threw his forces back into the race, he did it with flair.

First came Andrian Nikolayev, launched aboard Vostok 3 on August 11, 1962. Nikolayev's mission, according to Tass and Radio Moscow, was to gather information on man's abilities in space over a longer duration. How long he'd stay up there wasn't revealed. The Chief Designer's big surprise, his neat trick on a gullible world, came a day later. A new Vostok sat on the pad at Tyuratam. Nikolayev could see it belch smoke and fire as he came up on the launch site to enter his seventeenth orbit. What credit was truly due to Korolev was for demonstrating an ability to launch a space rocket exactly on time, perhaps to within a fraction of a second.

Vostok 4, carrying Pavel Popovich, followed almost exactly the same launch trajectory on August 12 that Nikolayev had followed a day earlier, getting into a slightly higher orbit that put him ahead of Nikolayev. That was another one in the credit column for Korolev. The rest of it was smoke and mirrors. Over the next day, Nikolayev's lower and thus faster orbit gradually closed the gap. On their eighteenth orbit, the two were just 3.2 miles apart. Radio Moscow hailed the fact, and the world concluded that the Russians had some kind of rendezvous capability in space.

Three miles is not a great distance. Nikolayev and Popovich could see each other, not in great detail but enough to see that the other Vostok was briefly there. This much was an interesting accomplishment. It wasn't all that difficult. It had the appearance of being another Soviet space spectacular, a first that proved Russia was still far ahead in the space race. The world's press played it that way. It seemed to be the first rendezvous in space, a vital precursor to going to the moon, to building a space station, to doing any of the big, important things out there. The Russians didn't *say* it was a rendezvous. They didn't lie or make claims that couldn't be substantiated. All they said was that Vostok 3 and Vostok 4 came within 3.2 miles of each other on the world's first tandem spaceflight.

TWO UPSMANSHIP, headlined a London newspaper. Sir Bernard Lovell of the Jodrell Bank Observatory gave the Chief Designer everything he wanted when he was quoted, "I think that the Russians are so far ahead in the technique of rocketry that the possibility of America catching up in the next decade is remote."

Dr. Edward Teller in California went a step further. "There is no doubt," he gushed, "that the best scientists as of this moment are not in the U.S., but in Moscow."

Wernher von Braun's more accurate assessment was at the bottom of stories, or ignored altogether. He saw no big advance in the tandem mission, he said, adding, "It does not look like the Russians used any new equipment."

They didn't. Nor was it like two aircraft flying together wingtip to wingtip, a formation called "station-keeping." Instead the immutable laws of physics took over. After Nikolayev's close approach to Popovich, the distance between them gradually grew to more than eighteen hundred miles, and neither of the cosmonauts could do a thing about it. When they returned to Earth, Nikolayev had stayed up four days, a new record, and Popovich was weightless for three days. Their retrofire maneuvers were planned carefully, and they landed on Earth, after ejecting and coming down the last leg under their parachutes, only six minutes apart.

They also landed 120 miles apart.

The launch timing did demonstrate that the Chief Designer's crews were getting better and better. But the most precise timing and control of a launch is never good enough to exactly duplicate another launch. There are too many variables, even including atmospheric density and winds aloft, which are never the same from day to day. If every variable was exactly identical, a difference in liftoff timing of even a millisecond is enough to ensure that two orbits will not be the same. That kind of timing is still, at the end of the twentieth century, so rare as to be impossible. Radar tracking data revealed the fake rendezvous gimmick to the U.S. military and to NASA. There was a sigh of relief that the Russians had not suddenly developed the hardware and the techniques needed to rendezvous. But there was frustration at the success they had in painting themselves as something they were not.

It was too late to call a press conference and to expose the magician's trick pulled off by the Chief Designer. Anything like that would have been called sour grapes. So NASA people issued polite congratulations, bit their collective tongues, and went on with the job.

The Russian flights put new pressure on the Kennedy administration. Mercury was flying, but there was a growing internal demand for presidential involvement, for some kind of public demonstration of Jack Kennedy's commitment. He decided to give it by making a two-day tour of space installations.

He started on September 11, 1962, at Cape Canaveral, where Launch Complex 34—the moon pads—and the Vertical Assembly Building, soon to be the largest building in the world, were under construction. The enormity of Apollo had to be obvious at that point. But it was driven home later in the day at Marshall Space Flight Center in Huntsville, Alabama, when von Braun led the way into a huge high-roofed manufacturing building and pointed to the first stage of a Saturn 1 rocket lying on its side. The rocket was twenty-two feet in diameter. It would be used in early Apollo tests, but the real moon rocket's first stage, with a thirty-three-foot diameter, was painted on a wall and dwarfed even this giant.

"This is the vehicle designed to fulfill your promise to put a man on the moon in this decade," von Braun told Kennedy. "And by God, we'll do it."

The entourage was in Houston the next day, and Bob Gilruth had Kennedy's ear about the Apollo spacecraft. He had Deke Slayton and Al Shepard standing by to take the president into mock-ups of the Apollo command module and lunar excursion. Kennedy grinned like a boy in a toy store when he wiggled the control sticks.

Then came the point of the whole trip. Jack Kennedy was determined to throw American superiority and determination in the face of the Russian bear. His speech at the Rice University football stadium had been billed as a major foreign policy and space declaration. More than fifty thousand people filled the seats to hear it. Kennedy didn't disappoint. The huge crowd threw out a cheer when he talked about the space race.

"We mean to be part of it," he said. "We mean to *lead* it, for the eyes of the world now look into space, to the moon and to the planets beyond, and we have vowed that we shall not see it governed by a hostile flag of conquest"—and now his voice was rising—"but by a banner of freedom and peace."

The cheers swelled and Kennedy paused under the hot September sun. When he went on, he rededicated the United States to supremacy in space. Then in a way that only this young president could pull off, in the tradition of his inaugural challenge to "ask not what your country can do for you; ask what you can do for your country," he put it all in a perspective that the NASA team fully understood and that he hoped the nation would adopt as a personal philosophy.

"We choose to go to the moon in this decade and to do the other things," he thundered, "not because they are easy, *but because they are hard.*"

Bob Gilruth would quote Kennedy on that often. It certainly was hard.

———

The race was run in fits and starts. Two months after the tandem Vostok flights, Wally Schirra took his Mercury capsule *Sigma 7* on a six-orbit flight, the longest time yet for an American in space but far short of Andrian Nikolayev's sixty-four orbits.

Schirra was a superb pilot, coming from barnstorming stock, and his touch on the Mercury hand controller was exquisite. He started out with the same amount of hydrogen peroxide thruster fuel as John Glenn and Scott Carpenter, did everything in the flight plan to perfection, and came home after nine hours, forty minutes in space with more fuel remaining than either of them. He did encounter a problem that plagued Scott Carpenter. Three times he inadvertently switched on both the automatic and the manual control systems simultaneously, consuming his fuel at a gluttonous rate. Each time he discovered the problem quickly, but reported, "It's much too easy to get into double authority." The switches simply were not well designed, and Mercury lacked the kind of computer ability that would eliminate such events on future spacecraft. But Schirra had the benefit of knowing what to watch for after Carpenter's debriefing and pointed reminders from Carpenter himself.

In the final twenty minutes of his mission, he controlled *Sigma 7* so delicately and precisely that he brought it down less than five miles from the aircraft carrier *Kearsarge* waiting to pick him up south of Hawaii. It was the first Pacific Ocean landing in the Mercury program. "I think they're gonna put me on the number three elevator," he joked after his main parachute opened and he got a good look at the *Kearsarge* out his window. On board the carrier, Schirra called it a textbook flight and NASA agreed. His performance helped to write the textbook for missions to come.

Some of the people who would fly those missions were getting their first taste of what it was like to be inside the action. A few weeks before Schirra lifted off, NASA brought in nine more astronauts to train for Gemini and Apollo. Two of them, Jim Lovell and Pete Conrad, had just missed when the Original Seven were picked. Two more were civilian test pilots, allowed to apply for the first time. They were Elliott See, a Navy veteran flying for General Electric at Edwards Air Force Base, and Neil Armstrong, the NASA test pilot at Edwards who'd learned that it didn't pay to oversleep.

The other five were military test pilots: Frank Borman, James McDivitt, Thomas P. Stafford, Ed White, and John Young. The Second Nine were a mixed lot, good pilots but not as good as the best of the seven. They were top-notch engineers, as a group better and smarter than the seven, but that was a close call. They were younger, had less combat flying experience, and less partying experience too. None of them would gain the skirt-chasing, hard-drinking reputations of six of the seven. Tom Stafford took a run at the record set by Al Shepard, but fell short. Pete Conrad and John Young were first-rate practical jokers, Young more on the laid-back, laconic side and Conrad out there pulling world-class pranks. The deacon of the group, the John Glenn act-alike, was Frank Borman, deeply religious and with a stern outlook that brooked no nonsense.

These Second Nine would write some of the most important chapters in the story of the space race, and two of them would die before another of them crossed the finish line. They were in a tough business. That was part of the challenge and part of the fun.

The weeks after Schirra's Mercury flight were not fun for Jack Kennedy or for the thousands of Reserve and National Guard troops called to active duty during the Cuban Missile Crisis. Nikita Khrushchev again was forcing a confrontation by putting intermediate-range attack missiles in Cuba. The Soviet Rocket Force was an expanding power, and the country's success in space had put an exclamation point on the image that Russian rockets were a potent threat. Khrushchev knew how to exploit his propaganda success and how to get what he really wanted. After Kennedy blockaded the sea-lanes around Cuba, turning back Soviet ships carrying additional missiles, the two countries came to a quiet agreement. Khrushchev removed missiles from Cuba under the full glare of international publicity. At the same time, Kennedy quietly removed missiles that had been installed in Turkey, near the Soviet border. Peace, such as it was in the simmering Cold War, continued.

Next up in Mercury was Gordon Cooper, the last of the Original Seven eligible to fly. His dry Oklahoma twang was irritating to some of the NASA headquarters people, which was increasingly being populated by bureaucrats far removed from the operational end of things. They saw Cooper as a yokel, ignoring his engineering and flying skills while concentrating on an accent they considered anything but cosmopolitan. It didn't help that Cooper was an independent cuss, given to racing fast boats on weekends, flying his own airplane during off hours, and generally turning up in newspaper photos doing something that might break a bone, or embarrass the bureaucrats, which was even worse. He compounded his hotshot image by doing an occasional low-level, high-speed buzz job past office windows at the Cape in a jet aircraft.

Even Walt Williams, long accustomed from his thirteen years in the desert at Edwards Air Force Base to the exuberance of men who flew dangerous machines, was getting disenchanted with Cooper. But astronaut boss Deke Slayton did what his boys expected: He stood up for Cooper against all the critics, reminding them that when they hired the Original Seven, they hadn't been looking for shoe clerks. Cooper eventually toned down his extracurricular activities, but not much.

Instead of rushing ahead to complete the Mercury program and get on with Gemini, Bob Gilruth's people took a deliberate approach to planning the final mission. At the same time, most of them were moving themselves and their families to Houston. Chris Kraft got a reminder that no one was immune from NASA's growing bureaucracy. After getting settled with his family in Friendswood, a small town southwest of the space center, he opened the mail to find a bill from NASA for $49.80. "This indebtedness was incurred due to your shipment of household goods over and above your authorized weight allowance in accordance with NASA Management Manual 26-13-1 P8," said the accompanying information sheet. Kraft sent the check on Valentine's Day 1963.

The Houston people knew there'd be a hiatus between Mercury and Gemini; they wanted to fly two unmanned Geminis on the Air Force Titan II booster before putting men on board, and those spacecraft wouldn't come off the assembly line at the McDonnell plant in St. Louis until late 1963 and early 1964. The earliest time they'd be ready to fly men would be late summer 1964.

Gemini was truly a spacecraft. The basic capsule looked much like Mercury, but bigger. Its top snout was longer and thicker too, to hold two rings of thrusters to be used for reentry control. Those small RCS jets—for reentry control system—were vital on every mission as astronauts plunged through the atmosphere. On one mission, they'd be the difference between life and death in a major orbital emergency.

Below the heat shield, Gemini and Mercury were very different. The Gemini ship added a hefty "adapter module," or cargo and supply container. First came a broad, slightly cone-shaped unit containing redundant retrorockets and their fuel tanks. Jet thrusters for controlling Gemini and maneuvering it in space were embedded in its outer skin. A second cone-shaped unit flared out from the retro carrier. This large cargo ring held fuel tanks for orbital maneuvering, rear-facing jet thrusters that gave Gemini true maneuverability in space and allowed it to do intricate rendezvous maneuvers, plus oxygen and water tanks for the crew, a fuel cell system that would combine oxygen and hydrogen to generate electricity and produce water, and other equipment.

Thus the overall shape of Gemini was conelike. Stand it on its base and it tapered upward, nearly twice as tall and twice as wide as a Mercury capsule. The crew compartment had two hatches that opened like car doors, and it did not have a rocket-powered escape tower. If something went terribly wrong on the pad or during the first minutes of launch, the astronauts were supposed to fire ejection seats that would throw them clear of danger, then bring them down under individual parachutes. The system was similar to the Vostok ejection seats, but angled so much forward that astronauts were afraid that an actual ejection might cause severe back injuries.

A few at NASA headquarters were ready to close down Mercury after Wally Schirra's flight, taking the program out on a near-perfect performance. The Mercury team in Houston disagreed. Another capsule was being readied on the assembly line in St. Louis, and Bob Gilruth's people were hatching plans to modify it for a seventeen-orbit, twenty-four-hour mission. The plan was approved, but one of the conditions was that it definitely be the end of Mercury. There was too much work to do on Gemini.

It took most of the fall of 1962 and into late winter 1963 to refit the capsule for a long-duration flight. The heavy periscope was removed, along with some redundant radios and data recorders. Water, oxygen bottles, additional coolant, and more hydrogen peroxide thruster fuel went on board. Then a plan to launch Gordo Cooper on his day-long flight in April went by the wayside when his Atlas rocket needed extra work to fix wiring problems. On the day it finally rolled out of the General Dynamics plant in San Diego, and had not a single recorded flaw, Cooper was there. He walked over to the rocket and used a marker pen to draw an arrow on its side, pointing up.

Next to the arrow, he printed, "Launch this way," and signed it, "G. Cooper, pilot."

Along the way, the Houston team saw that the *Faith 7* capsule and Cooper were good for even more time in space and increased the mission to twenty-two orbits, about thirty-four hours in space. After one launch scrub on May 14, Cooper rocketed into space the next day. He'd set some kind of record for calmness by falling asleep during the countdown on the first try and showing only a little increase in heartbeat and blood pressure as he rode toward space.

It wasn't quite a textbook flight. His sleep turned into catnaps when the system regulating his space suit temperature kept fluctuating. He'd get nicely asleep, then wake up either sweating or shivering. But he was getting his assignments done—photography of the Earth using a series of cameras, trying out various foods and learning to sip weightless water while wearing his helmet, spotting a bright light on the ground in South Africa. He jettisoned a tiny satellite with a strobe light, but it was hours before he spotted it drifting across Earth's horizon. An experiment to release a balloon failed. No one knew why the balloon didn't inflate after it was set loose.

It was in the last few hours that things went bad. First a light flashed on. It meant he was descending. He wasn't. Electrical problems were at work. On his next orbit, twenty down and two to go, key instruments died. Cooper couldn't read the attitude position of *Faith 7,* except by checking out the hatch window. An hour later his automatic stabilization and control system burned out. He was on manual control without instruments, an orbiting version of pilots in the 1920s and '30s who flew cross-country by watching the horizon and following roads.

Gordo Cooper could see the horizon, but he had no roads for landmarks. The ones he could see on the ground were no help, and some skeptical scientists didn't believe he could see them anyway. He could. Then something went wrong with the environmental control system. At the Zanzibar tracking station, Scott Carpenter took Cooper's lazy-style voice report on the latest trouble, then heard him follow it with a laid-back observation, "Things are beginning to stack up a little."

There was a pause before he went on in the same calm, twangy voice. "ASCS inverter is acting up," he reported. "And my CO_2 is going up in the suit. Partial pressure of O_2 is decreasing in the cabin. And my inverter won't come on the line." He gave Carpenter a moment to digest the information, then summed it up. "Other than that," he said, "things are fine."

John Glenn was at the Hawaii tracking station. When he heard from Cooper nearly a half hour later, at the far western horizon, there was still pure calm in the transmission. Cooper had used his manual controls and visual sighting on the horizon to set *Faith 7* up perfectly for retrofire. Glenn counted down and Cooper fired his retros manually, hitting the timing perfectly. "Have a cool reentry," Glenn radioed.

"Roger, John," Gordo Cooper replied. His tone of voice said it was just another day's work. "Thank you."

The *Kearsarge* was waiting just south of Midway Island. When the parachute opened over *Faith 7*, the second Mercury capsule to reenter and splash down in the Pacific Ocean, Cooper had made the best job of it yet. He'd been in space thirty-four hours, twenty minutes, and even flying manually during retrofire setup and reentry, he beat Wally Schirra's record by landing barely four miles off the carrier's bow.

Mercury was over. With it, America stopped taking those baby steps in space. The race was still running and the moon was a long way off.

9

What glory there was, a couple of parades and then back to work, didn't last long. Gordon Cooper got a ticker-tape parade in New York City and with his offbeat sense of humor, couldn't resist taking Trudy to a Broadway show: *Stop the World, I Want to Get Off.* Then they headed to Houston for the first astronaut homecoming parade in their new home. Six of the Original Seven bought houses in tiny towns and villages near the space center site at Clear Lake. Al Shepard opted for an apartment in a ritzy part of Houston.

Sergei Korolev had monitored preparations for the last Mercury flight with more than his usual interest. For Korolev's purposes, the open American program was ripe for his style of one-upmanship. So while Gordo Cooper and the Mercury team got themselves ready to close out Mercury, Korolev prepared to steal the headlines for the glory of Mother Russia. His plan might have been called Magic Act II. It was one part genuine accomplishment, one part slick showmanship. It began on the afternoon of June 14. Valery Bykovsky stood at the base of an R-7A rocket in his orange space suit, gave the now-traditional parting speech for cosmonauts, then was carried to the top by elevator and crawled into the Vostok 5 spacecraft. At exactly 3 p.m.—the Russians were getting good at countdowns and on-time launches—he was on his way to a not-quite-circular orbit 162 miles by 209 miles above Earth. His mission, Radio Moscow announced, would last up to eight days. Coming only a month after Gordo Cooper had maxed out Mercury's capabilities at thirty-four hours, it looked like the Russians were lengthening their lead in space.

Bykovsky's flight plan called for him to get detailed photos of Earth's horizon at sunset and sunrise, to take measurements of both the sun and the moon, and to carefully monitor radiation that passed into the Vostok capsule. Most important to Soviet flight surgeons, and to the spacecraft designers in Korolev's bureau who already were working on an advanced ship to be called Soyuz, Bykovsky loosened the straps holding him to his seat and floated free for ninety minutes each day. In the cramped compartment, he couldn't float far, but he was able to turn himself faceup or facedown, and he recorded his reactions to being free of restraints. Once he adjusted to weightlessness, the ability to free-float was not only enjoyable but showed how easily a man could adjust to a world where there was no up, no down, and a drawer could as easily be located in the ceiling as in a wall.

The long-duration flight would have been enough to convince the world that the Soviet mastery of space exceeded the Americans'. But Sergei Korolev wasn't done. Less than two days into Bykovsky's flight, another cosmonaut stood at the base of a rocket and gave a short speech. After a 12:30 p.m. launch as Bykovsky approached Baikonur on his thirty-first orbit, Valentina Tereshkova's Vostok 6 capsule passed 3.1 miles from Vostok 5. The first woman was in space and they had to hurry to make their joint announcement from orbit.

Tereshkova—they called her Valya—was an enigma and an anomaly. She was a worker in a fabrics factory and dabbled in sport parachute jumping. She was not a pilot, but the twenty-five-year-old was in strong physical condition. After Gherman Titov's flight, she wrote to the space bureau in Moscow volunteering her services. Nearly a year later, the letter forgotten, she was called for medical examinations, then taken into the cosmonaut corps along with several other female candidates. Sergei Korolev was thinking ahead, and putting a woman into space would be a major propaganda victory.

It also could be justified as a legitimate biomedical experiment. Until men flew in space, no one knew for sure how their minds or bodies would react. The same questions surrounded women, the "weaker sex," and to an even greater degree. In the United States, twenty women, including Jane Hart, the wife of a U.S. senator from Michigan, passed the full set of astronaut medical tests at the Lovelace Clinic in Albuquerque. But NASA was not ready to take women on board. Korolev saw his opportunity.

Tereshkova got barely a year of training, mostly more parachute jumping, along with a little copilot time in military transport aircraft. The medics put her through the intensive tests, ranging from recording her reactions to extreme heat and cold to isolating her in chambers of complete darkness. They knew Valya well as a physical specimen. Reports trickling out years later left doubts about her mental state.

Tereshkova's rocket didn't put her on the same precise trajectory they'd managed on the tandem flight a year earlier. Both her apogee and perigee were slightly higher than Bykovsky's, and she took about six seconds longer to complete one orbit. Still it was an accomplishment, and with both their faces being transmitted to Russian stations by onboard black-and-white television cameras, they told the world:

"Have started carrying out joint spaceflight. Dependable radio communications established between our ships. Are at close distance from each other. All systems in the ships are working excellently. Feeling well."

Newsweek speculated that the missions' true purpose was rendezvous and docking. By revealing only a few facts and letting Western journalists draw their own conclusions, the Russians were getting exactly what they wanted; if the conclusions were wrong, there was no need to correct them. In fact, the distance between the Vostoks quickly grew. In her higher orbit, Tereshkova fell farther and farther behind Bykovsky. It didn't matter. Nikita Khrushchev talked to her that evening on live television and used the moment to declare "the equality of men and women in our country."

A few American women reacted with both joy for what Tereshkova represented and anger against the United States for not allowing women to be astronauts. Jane Hart was quoted widely. "I'm more annoyed at the fact than I am impressed," she told one reporter. "It only showed that the U.S. is a hundred years behind in using the full abilities of women."

Even the chairman of the Senate Space Committee, Clinton Anderson, swallowed the Vostok story and joined in the NASA bashing. "The Russians have proved to us that you don't have to have twenty years of test pilot experience," he complained, "before you can handle one of these capsules." Like the rest of the world, he didn't know that Vostok had no provisions for "handling" anything.

The argument for women in space was valid then and has long been settled. But Mrs. Hart might have been less annoyed if she'd known what was happening aboard Vostok 6. Vasily Mishin, a Korolev deputy, told an interviewer in 1990 that Valya was close to coming undone during her flight. "Tereshkova turned out to be at the edge of psychological stability. It would seem that her flight . . . should have discredited N. S. Khrushchev."

Ivan Spitsa was a tracking site expert during the Vostok flights and said much the same thing in an oral history. "The poor woman was in such a state that her communications over the voice circuit were almost unintelligible," he recorded. "She variously became nauseated and went to sleep. We began switching her cabin light on and off. She woke and began calling out her code name: *Seagull here, Seagull here* . . ." At one point she was out of range of the Russian ground transmitters, but there was time to get a message to Bykovsky, who had a full-time radio link with Valya. Keep trying to wake her up, he was told.

None of that came out at the time. They brought Tereshkova down after three days, just forty-nine orbits. Bykovsky came down a few hours later, completing five days of his eight-day mission, and setting yet another space endurance record. He was in good shape and could have gone longer. The most plausible explanation for the early landings is that Tereshkova was sick and increasingly unstable. Once she was on her way, they had to bring Bykovsky down too, or face reporters' questions about terminating Valya's flight early.

The pair landed under their own parachutes several hundred miles apart. Safely back on Earth, Valya Tereshkova broke down into sobbing tears. Radio Moscow said she was overcome by her reception and by the flowers given to her by peasants. But she recovered rapidly and was an international heroine. She received congratulations from the queen of England, from American feminists, from women, and men too, the world over. A few months later she married cosmonaut Andrian Nikolayev and gave birth to a daughter the following year.

But in the light of what happened later, and what eventually became known, Valya Tereshkova's mental state might never have fully stabilized. In 1967, she told reporters that the Russian moon crew was already in training. Its commander, she said, was Yuri Gagarin. Then she added: "And I am on it."

It didn't quite work that way.

Did Jack Kennedy and Nikita Khrushchev have a cure for the Cold War? The Cuban Missile Crisis opened new channels between the American president and the Soviet premier, including a hot line directly connecting the White House with the Kremlin. The worst that could be said is that the two leaders were beginning to understand each other, though the understanding on each side was imperfect.

In autumn 1963, Kennedy renewed his proposal that Russia and the United States join forces in a lunar program. This time Khrushchev did not quickly say no. In another of their after-dinner Kremlin walks, Sergei Khrushchev asked his father what was different. "Now we have five hundred missiles, not two hundred," Premier Khrushchev said. "We can destroy five hundred American cities and that is enough to keep Kennedy from attacking us."

"Will you accept his offer on a moon program?"

"I don't know. I'm thinking about it."

Before Khrushchev made any public decision, Jack Kennedy was assassinated in Dallas. Whatever chance existed for the two to work together on a peaceful space program disappeared. Khrushchev may have said *nyet* again anyway. He'd already told reporters that the moon was not a Russian goal. "We are not at present planning flights by cosmonauts to the moon," he said in an interview that got little attention in the United States. "I have a report to the effect that the Americans want to land a man on the moon by 1970–80. Well, let's wish them success. We do not want to compete with the sending of people to the moon without careful preparation."

A year later Khrushchev was deposed by a militant faction of the Soviet Communist Party. The Cold War continued and got worse. The space race continued too, but one of the runners was getting weaker.

John Glenn was not a happy astronaut. After his orbital flight in 1962, NASA didn't give him much to do. He was a national hero, his face had been on magazine covers and newspaper front pages, and everybody knew his name. But with the Mercury program over and Gemini coming, Glenn had no firm assignment. Inside the astronaut office, it was quietly known that Al Shepard would get the first manned Gemini flight, with one of the Second Nine as his crewmate. Because of Chris Kraft's animosity, it was also assumed that Scott Carpenter wouldn't get any of the ten planned Gemini manned missions. The rest of the astronauts had hardware or mission-related assignments that were giving them influence and experience within Gemini and would probably lead to a spaceflight.

John Glenn had nothing. Every six weeks or so, he'd drop in on Bob Gilruth for a quiet conversation. Most of them ended the same way. "Bob," Glenn would say, "I'm fit and able, and I want to be assigned to Gemini."

Gilruth would give him a soulful look, sometimes rub the fringe of gray hair over his ears, and tell him the same thing for the umpteenth time: "John, I'm sorry, but headquarters just isn't ready for that yet."

What Gilruth didn't say, and perhaps didn't know, was that John Glenn had been grounded by no less than the president of the United States. Jack Kennedy had passed the word to NASA headquarters that John Glenn was a national asset, a treasure, and that he was not to be risked on any future space mission. Glenn would not learn of this until 1983. By then he had a successful second career going for him.

"I'd go see Bob and he'd tell me the same thing every time," Glenn said. "Eventually I got the message." When he did, he retired from the Marines in early 1964 and decided to run for the U.S. Senate from his native Ohio. That didn't work either. Early in the primary campaign, he slipped and fell in the bathroom, hitting his head and damaging the balance mechanism of his inner ear. He was forced to retreat to his home near the space center in Houston and live virtually as an invalid for long months while the damage slowly repaired itself.

It was a lonely and frustrating time. Test pilots have a lousy bedside manner, mostly out of superstition that whatever that guy's got might rub off. So Glenn had few visitors, now and then playing chess with a *Houston Chronicle* reporter who lived nearby, but mostly sitting alone in his terry-cloth bathrobe and being tended by Annie or trying to walk with the help of canes from room to room.

He slowly began to get better. But still he wondered: *Will I ever fly again?* He meant as a pilot. The idea of going back into space in a giant winged craft never crossed his mind.

Sergei Korolev's health was failing. The Chief Designer, now fifty-seven years old in 1964, was increasingly prone to pneumonia and other respiratory diseases. He suffered intermittent sharp pains in the lower abdomen, his heart was not strong, and he had headaches. None of that stopped him from plotting new and ingenious ways to do magic in space at the expense of the United States.

The openness of the American space program told Korolev everything he needed to know about Gemini. It was behind schedule by several months. The Titan II booster had needed modifications before the Gemini team, headed by Chuck Mathews, who'd helped design Mercury, would declare it man-rated. In a number of Air Force test launches, there were problems getting smooth combustion of the rocket's two-part propellants which were hypergolic—they exploded violently into fire simply by touching each other. The other problem was related to the combustion and to the lines carrying the propellants. On its way toward space, the Titan rocket jiggled up and down like some insane pogo stick. The vibrations sent shock waves through the rocket, enough to make the ride extremely uncomfortable for astronauts, and even do severe damage to the spacecraft itself.

The problems were fixed by refining the propellant nozzles and putting devices to smooth the flow of liquids in the propellant lines. The next step was to launch a Titan II with a dummy Gemini, and that finally happened in April 1964. When the spacecraft reached orbit with no pogo problems, both the Air Force and NASA declared that the Titan II was man-rated. One more launch was needed to man-rate the spacecraft. After that, men would fly in Gemini.

They'd do it without Walt Williams. Bob Gilruth's Manned Spacecraft Center empire was growing almost too fast. Williams was lobbying to have the center split into two components. Gilruth would have the spacecraft and astronaut training. Williams wanted a new center that handled operations; he'd be the czar of space missions. Gilruth didn't think the idea would fly and Williams lost his bid. He announced his decision to leave NASA and didn't even get to be operations director on the first Gemini flight. George Low of NASA headquarters sent a private note to Chris Kraft, explaining that he'd be in charge. "Both Gilruth and I feel that a 'lame duck' should not be put into the position of making the kinds of decisions involved in an operation," Low wrote. "Please do not discuss this note with anyone except Gilruth."

With the first unmanned Gemini a success, Williams took a position as a vice-president for the Aerospace Corporation in California. His departure solidified Chris Kraft's position as the number one flight director, and Kraft took over, too, as Gemini operations director. Before the year was out, he'd be elevated to a key position at the Manned Spacecraft Center as director of the Flight Operations Division. Kraft's career was ascending, but he knew that he'd learned much by watching Walt Williams make decisions. He sent Williams a letter from his heart. "It's beyond my comprehension that NASA would allow an individual of your talents and experience to leave when the jobs we have to do are so staggering," he wrote. ". . . Doing it without you seems to me to be an almost impossible task."

Originally Grissom and Young were the backup crew for Gemini 3. Al Shepard was first given the mission, with Tom Stafford as his number two. But fate took a hand in things. Shepard began experiencing dizzy spells, sometimes with nausea and even vomiting. He kept it hidden for a while, but it got worse. The symptoms were the same as John Glenn's, but Shepard hadn't suffered a fall or any other head injury. When it got worse, he went to see Dr. Charles Berry, head of the Manned Spacecraft Center's medical office and unofficially known as the astronauts' doctor.

The news was as bad as it gets for a pilot. Shepard had Ménière's syndrome, an inner-ear infection that disrupts balance functions. Medication could mostly control it, and it might—a long shot—go away with time. There was no known sure cure. Meanwhile Shepard was grounded. He couldn't fly airplanes solo. Deke Slayton commiserated—he knew exactly how his friend felt—and removed him from his assignment to Gemini 3. There was a consolation prize. Gilruth appointed Slayton, whose managerial and administrative skills had blossomed, to head a new Flight Crew Operations Division (FCOD) at the Manned Spacecraft Center. FCOD would be responsible for everything involving astronauts—their training, their aircraft, their day-to-day work assignments, their mission assignments, even the well-equipped private gym with its own outdoor running track located near the north boundary of the space center.

When Slayton moved up to the director's level, he left a void, but he knew how to fill it. He handed Al Shepard responsibility for running the astronaut office. Another fourteen astronauts had been signed on in the spring of 1964, bringing the total to thirty including the hopeful lame ducks John Glenn and Scott Carpenter. Shepard took the job, but not happily. He wanted to fly, and being chief astronaut wasn't even second best. His mood swings showed his displeasure. He could be charming one minute, sarcastic and biting the next. He got the job done and was the astronauts' mother hen, but having him as a boss was not often a pleasant experience.

Shepard's secretary was a bright and experienced administrator named Gaye Alford. It didn't take long for her to get tired of astronauts sticking their heads through her door and asking what kind of mood the boss was in today. She called the space center's photo department and they dug up two mug shots of Al Shepard, one smiling and one scowling. She'd gauge his mood when he arrived each morning, then post the appropriate photo in a frame on the door. Shepard was no dummy; he knew what she was doing and why. But he didn't tell her to stop.

Eventually Slayton got enough complaints from the guys about Shepard's bad days. He called his old friend on the carpet, reminding him of the reality of both their situations. It helped. The scowling Shepard mug shot stayed in Gaye Alford's desk drawer more often than not.

The Americans in charge of putting a man on the moon seemed almost too confident in 1964. But NASA officials kept close counsel about their perceived realities of the space race. In congressional testimony, and in answer to frequent questions from reporters, the answer about a lunar landing was always the same: "We'll meet Kennedy's deadline. We'll put men on the moon before the end of the decade."

They had their reasons. A declassified 1964 planning document lists manned Earth-orbital missions starting in 1966 and lunar missions in 1967 and 1968. On the plan's date, May 7, 1964, the timeline still looked possible.

Voskhod was another of the Chief Designer's smoke-and-mirrors space programs. Sergei Korolev intended to upstage the American Gemini program before it got off the ground. He was smart, but any look at Korolev has to show that he also was lucky. Voskhod was his roll of the celestial dice, and Korolev came up a winner one more time.

If the Americans can put two men in a Gemini spacecraft, then Korolev saw no reason why he couldn't raise the limit to three. But he had a problem. A multimanned spacecraft called Soyuz was still in its early development stage. No Soyuz would be ready to fly until 1967. By then, the entire Gemini program would be history and the Americans would be flying their three-man Apollos. He took the direct approach. With the bitter 1964 Russian winter still howling, the Chief Designer put his engineers and technicians to work on figuring out how to turn a one-man Vostok capsule into a three-man Voskhod. It wasn't easy, but from competing designs Korolev chose one proposed by Konstantin Feoktistov. Then perhaps to ensure that only the best effort went into the conversion, he went one step further. He told Feoktistov that he'd be one of the three to fly it.

The Vostok capsule was comfortable for one cosmonaut in a space suit. For three men, it would be misery. Feoktistov's proposal started with eliminating the bulky ejection seat. That carried grave implications. The cosmonaut's personal parachute was part of the seat apparatus. Without the seats, the three cosmonauts would have no way to eject; they'd have to ride Voskhod all the way to the ground. They all knew that Vostok hit hard, even under its own parachute. They solved that problem by designing a pair of retrorockets that would fire just before the spacecraft hit hard ground. That would cushion the touchdown, though it would still be a jolting landing. A second parachute was added. That would help.

There was still no room for three men to wedge themselves into the capsule. So they eliminated the space suits. Any major leak in the pressure vessel during launch, orbit, or most of reentry could now have only one result. The cosmonauts would die. Korolev's team worked doubly hard to make Voskhod airtight.

It was still a tight fit for three men. There was no question of setting any kind of long-duration space record. The cosmonauts would be lucky to avoid muscle cramping even on a short flight. No matter. The first Voskhod had only one purpose. It would look good to the world; it would upstage Gemini and be in space months before the Americans were ready to go again. They built a test vehicle along with the primary capsule and launched it under the name of Kosmos 47 on October 6, 1964. It stayed in orbit one day and landed safely. American analysts noted its orbit and reentry, then logged it in as a "probable manned precursor."

Korolev didn't wait. With the Russian winter fast approaching, weather became a major factor in scheduling launches. He had a rocket and another capsule ready almost immediately. The three cosmonauts—Vladimir Komarov, a pilot; Feoktistov, the engineer; and Boris Yegorov, a medical doctor—crammed themselves shoulder-to-shoulder-to-shoulder into Voskhod on October 12 and were launched into an egg-shaped orbit with a peak altitude of 254 miles. The view from there was spectacular, at least what they could see of it by craning their necks to get a glimpse through the window. They didn't have much else to do; this mission was mostly a ride-along.

Feoktistov was the first nonpilot engineer in space. Yegorov was the first doctor. Along with being members of the first three-man crew, that got them into the record books. Yegorov brought along some simple medical experiments, checking vestibular functions in zero gravity, drawing blood samples, and measuring his comrades' eyes for light sensitivity.

But with Feoktistov almost sitting in his lap, there wasn't much else he could do.

For the next twenty-four hours, the world's press played the Voskhod story on page one. True to Korolev's prediction, everyone assumed that the Soviet Union had a new, three-man spacecraft and that it was again demonstrating a vast lead over the United States. The three came down after only a single day in space. The twin parachutes and retrorocket worked, but they emerged from the capsule stiff, sore, and only vaguely triumphant. It had not been a great trip. But Korolev and to a lesser extent Nikita Khrushchev got what they wanted. If they were now secretly lagging on the technology front, the Soviet program appeared to be far ahead in the space race. Image was everything.

It didn't do Khrushchev much good. The day after Voskhod 1 landed, the Central Committee of the Soviet Union deposed him. The new premier was Aleksei Kosygin, but the real power sat with Communist Party First Secretary Leonid Brezhnev, a hard-liner who disdained Khrushchev's softening toward the West. Korolev shrugged and went on. Brezhnev and Khrushchev had one thing in common: They both understood the value of smoke and mirrors.

An astronaut died and the author was pulled into the aftermath. It was a Saturday morning, Halloween Day 1964. Somehow my city editor at the *Houston Chronicle* got Deke Slayton on the phone, then reached out for me.

"It was Ted Freeman," Les Bennet said. "Deke's on his way to inform the wife. He said for you to meet him there. I promised him that we wouldn't go with the story until the wife knows. You see Slayton and call me immediately. We're already holding the presses."

I never knew how he'd reached Slayton; he wouldn't say, and as events unfolded, neither would Slayton. Ted Freeman was one of the New 14, an Air Force test pilot who'd just settled into a house in the Nassau Bay subdivision across the highway from the space center. He'd been on a routine training flight in one of the NASA-owned Northrop T-38 twin-seat jets on that Saturday morning. The T-38 was a high-performance trainer with supersonic capability. In another guise, virtually the same aircraft was the F-5 fighter-bomber that would soon bring devastation to Vietcong villages and ground forces in Vietnam. The T-38 was a touchy but forgiving jet. Astronauts loved it and generations of Air Force pilots trained in it. Somehow Freeman's T-38 had crashed and he was dead only six miles from his house.

I was twenty-four, an Army veteran, and I'd seen violent death in the service and on a brief tour as a police reporter, but the idea of being anywhere near Ted Freeman's house sent shivers through me. I lingered in the space center newsroom, the only reporter who knew the story, drank their coffee, then made the half-mile drive from the space center to Freeman's house. There was a silver 1964 Corvette parked at the curb. Deke Slayton drove a silver 1964 Corvette. Many of the astronauts had Corvettes, thanks to the sweetheart lease deal with Chevrolet.

When Slayton failed to appear, I walked to the door and rang the bell. Faith Freeman answered and I froze, then stammered, "Uh, is Deke here?"

Faith Freeman knew me, and in an instant she understood. Still, she had to ask. "No. Why should Deke be here?" But the look on her face said it all. She knew.

"Uh . . . there's been an accident. Deke's supposed to be here."

She was a strong woman, a test pilot's wife. All of them had sat with wives whose husbands would never come home. She knew the drill. That didn't make it any easier.

"Why don't you wait right here," she said gravely. "I'll go make a few calls and find out what's going on." She closed the door and was gone. It opened a half minute later, a very large and angry neighbor coming outside intent on doing havoc to what he perceived as a callous young reporter.

The silver Corvette at the curb belonged to the astronauts' physician, Dr. Charles Berry. He'd arrived, found that Deke Slayton was off somewhere procrastinating, and had gone to the neighbor next door. At that moment Slayton pulled to the curb. He'd included a stop at the crash site and a second stop at a bar for a quick jigger of courage. His Corvette and Doc Berry's were identical.

"Yeah, I'm late," Slayton growled at me, waving the angry neighbor away. "Go make your call."

When he found that Faith Freeman had already figured things out, Slayton's anger erupted. He buttonholed Paul Haney, now head of public affairs at the Houston center, and demanded that I be arrested.

"For what?" Haney asked. "He was doing his job. You're the one who said for him to meet you there."

Slayton fumed and dropped the whole subject. But it festered until two years later when I was working for both *Time* and *Life* magazines and Neil Armstrong banned me from an interview. No way, Armstrong said, was he going to talk to that reporter who got his kicks by telling widows that their husbands were dead. Haney and I went to Slayton, who'd forgotten the unfortunate circumstances.

"You're right," Slayton said. "It was my fault. I'd never delivered that kind of message and I took my time getting up the courage to go to the house. I'll talk to Neil."

He did. Neil Armstrong and I made peace. Nothing similar ever recurred. In the wake of Ted Freeman's death, Slayton set up quick and rigorous procedures to send astronauts to inform and counsel families in the moments after a tragedy. They'd be needed.

Ted Freeman died from a tragic combination of bad luck and too much courage. His T-38 ran through a flock of geese at five thousand feet. Both jet engines flamed out when goose parts were ingested, and at least one goose smashed the T-38 canopy. Freeman reported the problem to the tower at Ellington Air Force Base, just north of the space center. They told him to punch out, save himself, and let the plane go down in an open field.

But Freeman thought he could dead-stick the plane back to Ellington. He'd once saved another jet in a flameout landing and he could do it again. He couldn't.

The T-38 was lined up for the runway when Freeman knew that his descent rate was too fast. The plane would go into buildings under the approach path. He saw a field ahead and banked to drop the T-38 there. Then he ejected from the crippled plane.

Ted Freeman went out almost sideways at just one hundred feet altitude. His parachute was only beginning to open when he hit the ground.

The smoke was thinning and the mirrors were cracking. But the Chief Designer had one huge headline grabber left in his bag of tricks. While the Feoktistov group worked on the three-man Voskhod, a second group of Sergei Korolev's engineers converted a pair of Vostok capsules into still another Voskhod hybrid. This was the mission that people would remember. If anyone had seen the capsule in orbit, in all its improbable glory, they'd remember that forever too.

One more time, the Chief Designer set out to bring praise to the Russian space program by beating the Americans. Gemini was designed from the drawing board onward so astronauts could go outside on space walks. They'd learn to work while floating free in their bulky space suits, and they'd be trained to make repairs or to move from one craft to another. That might be needed in an Apollo emergency. In Gemini, the other craft would be an Agena rocket, not a ship that could be entered. But learning the techniques of moving from here to there was what mattered.

Korolev wasn't concerned about his cosmonauts' learning curve. The experience would be a bonus. He understood that the first space walk would be historic. That was enough to convince the bureaucrats to let him proceed. Voskhod 2 had only one purpose: do that space walk before any Gemini flew with astronauts.

The capsule Korolev produced was almost phallic. A small circular hole, just wide enough to let a cosmonaut wiggle through, was cut into the pressure shell and fitted with a removable hatch. Outside the hole, engineers installed a special metal ring. To that they attached a collapsible tube of a rubberized canvas-like material with struts and bands. It would deploy to a stiffened length of more than six feet from the spherical capsule. The tube had a second hatch at its far end, making it into the world's first air lock in space. After the space walk, the tubular contrivance was to be jettisoned.

It had to be tested in space, so two of the special Voskhods were built. Korolev was coming down with another case of pneumonia, but was in the control center when they launched the test capsule on February 22, 1965. It reached orbit and the air lock deployed on schedule. Then the test Voskhod passed out of range after receiving a series of radio commands from the ground and was never seen again.

The solution to the mystery should have given every cosmonaut who'd flown, and those who would fly this mission, a bad case of the nerves. Sergei Korolev made it a point to give his cosmonauts the extra three numbers they'd need to manually activate the retrorockets through the six-number code panel. That trust must have made each of them feel good and instilled them with added confidence. But did Korolev also tell them that the Vostoks and Voskhods carried hidden explosive packages so that controllers could destroy a capsule coming down in enemy territory? The fact of the explosives was revealed to the public in a 1993 interview of Soviet space suit expert Gai Ilyich Severin by American author James Harford.

The explosives, it seemed, were set off by conflicting radio commands.

There was no choice except to declare the mission a failure. Again nothing in space got past the Americans, who noted the launch and recorded it as "exploded; probable manned precursor."

Now the Chief Designer had a problem. The Americans had successfully launched their second unmanned Gemini a month ago, in January, after a series of delays caused by incoming hurricanes, a lightning strike on the rocket, and one abortive launch that saw the Titan's engines shut down one second after firing. In the January launch, the spacecraft worked perfectly, arcing up to space, then slamming into the atmosphere to prove that it could withstand the maximum speeds and temperatures it would ever see during reentry. The decision to go with men on the third flight was easy.

GT-3, with Gus Grissom and John Young aboard, was on schedule for a quick three-orbit trip in late March. No space walk was planned, but the Americans were determined to fly a new Gemini mission every two months for the next year and a half. If Korolev took the time to build and test another Voskhod, that historic first spacewalker would almost certainly be an American, not a Russian.

He couldn't let that happen. Korolev picked the crew on March 15, from several cosmonauts who'd been training for the mission. Pavel Belyayev and Aleksei Leonov were in space just three days later. Leonov's space suit was specially designed for spacewalking. One of its features—aside from the obvious—would soon save his life.

Unlike the pure oxygen atmosphere of American spacecraft, the air in Russian space capsules was exactly that: air. Soon after they reached orbit and deployed the air lock tube, Leonov struggled into a backpack apparatus containing pressurized oxygen. As they coursed around the world, he breathed the oxygen and it cleaned the bends-causing nitrogen from his blood. On their next pass over Soviet tracking stations, Belyayev opened the air lock hatch and, Leonov reported later, "I swam into the chamber."

While Leonov clipped a lifeline to his suit, Belyayev closed the hatch and opened valves to exhaust the air lock's atmosphere into space. He flipped the electrical circuit that opened the outer hatch, and Leonov found himself in "a blinding ray of sunlight." He waited until Belyayev gave the "go," then pushed out of the air lock and drifted free. He pulled himself back and removed the cover from a camera that caught him in its lens and transmitted the pictures to Russia. A Russian was walking in space! Korolev had done it again!

Leonov spent only ten minutes out there, tugging on his lifeline to return to Voskhod, pushing free again, pulling himself back. "I . . . stretched out my arms and legs and soared," he reported. Then it was time to return. But he couldn't. His space suit had ballooned in the past minutes and he was too big to reenter the air lock. He struggled and pulled. It was no good. Safety and survival were six feet away and unreachable. In his postflight report, the parts made public, he glossed over the problem. "I thought a bit, mentally pulled myself together and hit on a way of doing it quicker," he wrote. That was all Korolev released for public consumption.

Leonov was near the edge of panic, but did pull himself together. The quicker way also was dangerous, but he knew that there only was one choice. He reached down to a pressure valve on his special suit and began slowly letting out high-pressure oxygen. The suit shrank a bit, but quickly began to swell again as his backpack oxygen unit responded. It took him twelve long minutes to get it right. Bleed out oxygen, move deeper into the air lock, bleed out more oxygen, crawl a bit more . . . Finally he was inside. Belyayev triggered the outer hatch closed and repressurized the air lock. Leonov was safe.

He clawed his way to his seat, removed his helmet, and took a deep breath. "I felt streams of perspiration running down my forehead and cheeks," he wrote. "I think that it is a bit too early to compare outer space with a place for an entertaining stroll . . ."

The adventure of Belyayev and Leonov wasn't over. This Voskhod was built in a matter of months. When reentry time came after one day in space, its retrorocket system was dead. There was only one hope. Flight controllers rapidly calculated the angles and told Belyayev to set up for a manual reentry using one of the two landing rockets that were supposed to cushion their touchdown. It was tricky and uncertain. It was all they had. Voskhod 2 stayed in orbit an extra ninety minutes while Belyayev put the craft into position and Leonov checked his numbers. That wasn't enough time. They missed the countdown time by enough to send them more than twelve hundred miles farther downrange than their rescuers expected. Then it got worse.

The rocket brought them down from space. The parachutes opened. The single landing rocket remaining fired perfectly. But they were coming down not far from the city of Perm, in the Russian *taiga,* the thick northern forest of evergreens and birch. Voskhod 2 plummeted through thick upper boughs and landed in snow. It was wedged tightly between two tree trunks. From there, some details of the story vary. Certainly the cosmonauts stayed in the capsule. Either the hatch was held tight by a tree, or they opened it slightly. It was bitter cold outside, and as night fell, wolves howled and sniffed this strange contraption. Belyayev and Leonov were not injured, but they couldn't move much either.

They were found within hours of landing, and warm clothes and rations were dropped by helicopter. The *taiga* was too dense for any landing attempt. The relief packet lodged in the treetops, out of reach. But perhaps the hatch was jammed shut anyway. It was a miserable night.

By morning they'd gotten the hatch open and crawled out. A helicopter crew reported that they were chopping wood and had a fire going. But the rescuers couldn't land. Sergei Korolev sent in another helicopter, and an Alpine rescue team went down rope ladders into a narrow uphill clearing. It was late in the afternoon when they skied down to the cosmonauts with more warm clothes, food, tents, and extra skis. They stayed again overnight in the *taiga,* but warmer and more comfortable in down sleeping bags. On the morning of the second day, all of them skied twelve miles to a big clearing where the helicopters waited.

The world's first spacewalker finally made it home. The Soviet press machine hailed Belyayev as the first cosmonaut to manually control reentry from space. The stories didn't mention why it was necessary. In the United States, *Time* was discerning enough to question whether it was "forced by some failure," but *Newsweek* didn't even get the landing right. It went with the initial Radio Moscow report that Voskhod 2 landed at the "pre-assigned spot" on a lush state farm. Still, this Voskhod mission contained clues about how the space race was really progressing. Some Americans were quietly saying that the Russians were out of it. When their musings were printed at all, they were attributed to nameless "space experts" and were buried within stories that kept the race alive and its winner in question. Rumors would soon circulate in Moscow that a space station, not the moon, was the real Russian goal.

Two comments in the days after Leonov's space walk might have been given more weight if the press had truly been paying attention. The first was broadcast on Radio Moscow, by Vasily Seleznev, identified as a Soviet space official. Leonov's space walk, Seleznev said, "was made with a prospect in view that in the future cosmonauts will take part in assembling spaceships."

Then the Chief Designer himself, still nameless in public, issued a statement. "We shall yet live to see the day when orbiting platforms appear in space," he said, "resembling scientific research institutes in the Earth's upper atmosphere."

Sergei Korolev had insights that Bob Gilruth didn't. He knew what the Americans were planning in Gemini, and that Wernher von Braun's Saturn rockets for Apollo would certainly begin their test launches in 1966 or 1967. He also knew exactly the condition and prospects of his own space program. He was already facing internal competition from other design bureaus, and the people arguing for stations in space were getting some ear-time from Leonid Brezhnev and Aleksei Kosygin.

Most important to Korolev, he knew that he'd completely emptied his bag of tricks with the Voskhod 2 mission. He'd had other missions planned, all gimmicks with crews including a poet and a journalist. But the problems with Voskhod 2 were significant and needed work if more Voskhods were to fly. Given what the Americans had coming, there didn't seem to be a point. No matter who flew aboard a Voskhod, he'd still be just a passenger and he'd be adding nothing to the techniques of spaceflight. Voskhod with a poet-cosmonaut would look pale compared to a Gemini doing a legitimate space rendezvous.

So there would be no more Voskhods, though the Americans—space experts and press alike—speculated that this "new program" was just beginning. The next manned spacecraft by Korolev was still being developed. It would carry two or three cosmonauts more comfortably, and it would finally give them the maneuvering ability to match Gemini in orbit.

But a manned Soyuz was still two years off. The first manned Gemini flight was only days away. If the Americans were either lucky or good, or both, the entire Gemini program would be completed before the next cosmonaut flew in space. Korolev could only be content that he'd put Aleksei Leonov in the record book and his face on magazine covers and his name in headlines around the world. There was nothing Korolev could do about Gemini. For now, Comrade Brezhnev was happy. The Americans looked like losers again. So Korolev was content. He canceled his Voskhod plans and went back to work on Soyuz and maybe the moon.

If he'd seen the coming darkness, the Chief Designer might have felt differently about the space race and how he'd run in it. Or maybe he never had a choice.

10

In the twenty-two months between Mercury and Gemini, the United States began to lose some of its enthusiasm for the space race. The Voskhod missions rekindled it, but some damage already had been done.

Lyndon Johnson was less friendly to NASA as president than he'd been as vice president. A week after Jack Kennedy was assassinated, LBJ signed an executive order renaming NASA's launch facility as the Kennedy Space Center. At the same time, he ordered the entire Cape Canaveral area to be renamed Cape Kennedy. That latter decision was not popular with locals; Canaveral was a name going back almost to the discovery and first charting of Florida. President Nixon would quietly give Canaveral back its historic name in 1973, but Kennedy's name stayed attached to the space center.

At the same time, LBJ was dealing with an escalating war in Southeast Asia, with pushing through major civil rights and voting legislation, and with expensive social programs under the general name of the Great Society. Student unrest and an increasingly violent reaction to the war consumed LBJ. NASA was becoming a national luxury and President Johnson did it no favors.

The 1965 federal budget was the first fully controlled by LBJ. NASA's increase of $150 million, to $5.25 billion, was less than it asked for, less than it needed, and less than the national inflation rate. It was the only increase LBJ granted during his presidency, and even that was a decrease when inflation was factored in. With more missions than ever to fly in the Gemini program, Apollo to come, and manned flight planners hoping to capitalize on space successes by moving onward after Apollo to space stations and planetary missions, NASA thereafter saw its budgets under LBJ chopped mercilessly.

"Requirements of Vietnam and the Great Society have pared NASA's fiscal year 1967 budget request to $5.012 billion . . . putting off approval of a follow-on manned flight program for another year," the magazine *Missiles and Rockets* reported in January 1966. Five months later the approved budget was $4.986 billion, and soon NASA administrator Jim Webb was saying "A $5-billion budget level in the years ahead *will not be adequate . . .*"

Win or lose in the space race, Lyndon Johnson did some handwriting on NASA's wall, and the agency's hierarchy knew it. His early rhetoric had turned to empty words, and at the end, he was no friend of NASA. For fiscal year 1969, LBJ's last budget, the space agency got $3.991 billion, which wasn't even adequate to do more than think wistfully about the question "After Apollo, what?" No one knew.

Astronaut John Young felt something about the Russian space program. It was envy, mostly that they had flown a three-man crew in a "shirtsleeve environment." It just seemed so much more comfortable than the bulky space suits he and Gus Grissom had to wear, and his comments to the press proved to Sergei Korolev that the image of a good flight still overcame the reality of how it came about.

Grissom was more realistic. He'd read Yuri Gagarin's memoir, *Road to the Stars,* and correctly deduced that cosmonauts were just passengers who didn't participate in the design or testing of their capsules. That didn't please him at all. Through Mercury, and now Gemini, he was used to having a say in spacecraft development. He and the rest were test pilots, he'd say, and "we knew what we were talking about." The flight of Gemini 3, Grissom told reporters, was nothing more than a test flight. He'd been a test pilot since coming back from one hundred combat missions in Korea. John Young didn't have the combat time, but he was an experienced test pilot and he'd held some world flying records in Navy jets. Neither of them was a talker. They were jokers. Grissom was still smarting from four-year-old press criticism over the sinking of *Liberty Bell 7.* He decided to make a bold statement this time around.

He was a bit too bold for the growing number of humorless bureaucrats populating the offices at NASA headquarters in Washington. When he named the Gemini 3 spacecraft *Molly Brown,* a wry tribute to the unsinkable of the same name, the bureaucrats went red in the face. Hot phone calls and memos were mostly one-way, from Washington to Houston. Manned Spacecraft Center director Bob Gilruth backed Grissom and the name stayed.

But headquarters knew how to play that game. It issued an order that after Gemini 3, spacecraft would not have names. Future Geminis would be known as Gemini IV, Gemini V, on out to Gemini XII, with Roman numerals, not Arabic numbers. The stiff-necks in Washington took some of the fun out of Gemini, but they didn't care. Image was everything. On Gemini 3, image took a beating.

Launch was scheduled for 9 a.m., March 23, 1965. One of the tests on the test flight started the night before at Wolfie's, a deli restaurant by the Ramada Inn in Cocoa Beach and an all-night hangout for reporters looking for sober-up coffee, space people looking for food and talk, and even an occasional astronaut. The astronaut after dark on March 22 was Wally Schirra, picking up the tightest and most compressed corned beef sandwich that Wolfie's could put together.

Molly Brown lifted off the next morning after a twenty-four-minute delay. It was the least ambitious of all the Gemini flights—three quick orbits, a couple of significant maneuvering tests, activate some medical and biological experiment packages, and come on home. Engineers and flight planners in Houston were most anxious for the orbital maneuvers and for confirmation that the Gemini spacecraft itself was functional and at least moderately comfortable. On those two factors rested the future of the program and of key parts of Apollo.

Gemini had bigger built-in thruster rockets and carried considerably more fuel than Mercury, Vostok or Voskhod. It was designed to do extensive orbital maneuvers. When *Molly Brown* separated from the Titan's second stage, Gus Grissom's first job was to fire the rear thrusters and give the spacecraft its final boost into orbit. Watching his instruments, he held on just a little longer than planned and it was a good thing. Even with the extra thrust, *Molly Brown*'s orbit was just barely low, 76 miles by 109 miles instead of 76 by 113.

That low perigee was part of the plan; it was to be the shortest space trip by either the United States or Russia since the flights of John Glenn and Scott Carpenter. They didn't need a high ride into space. By the time they passed back over the Cape, Grissom and Young had broken one biology experiment with a balky handle and managed to activate a few others. Young spotted some disturbing readings on instruments, particularly one that said they were losing oxygen pressure. It took him less than a minute to identify a bad power unit, switch to a backup, and get readings that were normal again. Then it was Grissom's turn to begin exercising *Molly Brown.* He turned the spacecraft around and for seventy-five seconds fired Gemini's biggest thrusters directly into the line of flight. The maneuver slowed *Molly Brown* by fifty feet per second, enough to drop its apogee and make the orbit nearly circular at seventy-six miles. Radar tracking showed that he got it just right.

Now came Gemini 3's most important maneuvering test. Grissom turned *Molly Brown* ninety degrees and fired the jets again. True to calculations, the spacecraft moved a full mile sideways in orbit; it had gotten off the original space track and moved onto a new one. The change was only a fraction of Gemini's capabilities, but it was enough. In that moment, ground controllers knew that astronauts could guide a spacecraft to rendezvous with another, even if their orbits didn't at first perfectly match.

"Gemini's a Corvette," Grissom said later. "Mercury was a Volkswagen."

They had a few minutes before the next maneuver, and it was time to sample the freeze-dried and compressed space food packed on board. Grissom and Young didn't need a meal, but the medics and the food experts wanted some feedback on the taste of the food in orbit and whether it was easy enough to eat. "Hey, Gus," Young asked, "how'd you like corned beef about now?"

"Sure, sounds good," Grissom grunted as he worked to open one of the space food packs.

That was when Young pulled out the corned beef sandwich he'd gotten from Wally Schirra and smuggled into space: "Okay, try this one." Grissom laughed and took a bite. It was good, so he took another. Then he saw a bread crumb floating free and quickly repacked the sandwich. Food particles in zero gravity could have a way of getting into switches or instruments and causing problems. It was a good joke, but Grissom and Young both knew when to quit.

It was too bad for both of them, and for NASA, that some congressional critics didn't have the same restraint. A few representatives seized on the little joke and blasted NASA for not being able to control its astronauts and even for disrupting medical evaluations of their postflight condition. That fallout would come in the days after the mission ended.

With space food nibbled and the sandwich packed away, Grissom did one last big maneuver with *Molly Brown*. Turning backward again, he fired its jets until his instruments told that their orbit had dropped to a low point of just forty-five miles. The next time around, they'd reenter to a landing whether or not the craft's retrorockets worked. There were five of them, and after jettisoning Gemini's adapter section, Grissom fired them too. They were a bit too powerful. *Molly Brown* came down on the right track, dangling under its parachute, but it splashed fifty-two miles short of the carrier *Intrepid*. Until Grissom released the parachute, the spacecraft was being dragged nose-down through the water. Then it bobbed upright and proved, as John Young reported, that it "was no boat."

Grissom adamantly refused to open the doorlike hatches. He wasn't going to see another spacecraft sink under him. But it took a half hour for the rescue teams to arrive. The bobbing and wobbling of ocean swells, and the rising cabin temperatures, combined to show the difference between the Air Force man Grissom and the Navy man Young. Grissom inelegantly vomited, breakfast, corned beef, and space food all coming up. Young knotted his muscles and kept his stomach relatively calm. It was a minor glitch in a perfect mission. Gus Grissom and John Young proved that Gemini was everything it was supposed to be, except perhaps a good boat. Their mission gave a green light to the ambitious flights ahead.

Still, the Corned Beef Caper came back to haunt them. They were invited to the White House, where President Johnson added a cluster to Grissom's NASA Distinguished Service Medal. Young was pointedly ignored and received no medal at all. Nor did he get the military promotion that now was customary after an astronaut's first spaceflight. That would have to wait until he flew again a year later. Wally Schirra had bought the sandwich. But John Young paid the price.

A few days later Deke Slayton, who had known about the sandwich joke in advance, issued an order that no bootleg be carried on board a spacecraft without his approval. It was an interesting way to phrase it and Slayton-approved bootleg continued to fly, but quietly.

The world was listening and Ed White was having a ball. There was no television camera aboard Gemini, so photos of the first American space walk had to wait until they came down. But the sounds over the radio were exciting enough. It was the first U.S. spaceflight without one of the Original

Seven on board. Both mission commander James A. McDivitt and his co-pilot, Edward H. White, Jr., came to NASA as part of the Second Nine. John Young was the first of their group to fly. Now two more were up there and they were making history.

A space walk had been on Gemini IV's "maybe" list since the previous summer. When Aleksei Leonov went outside Voskhod 2 in March, it quickly moved to a "must" for McDivitt and White's four-day flight in early June. After watching video of Leonov seemingly having trouble controlling his movements in space, engineers at the Manned Spacecraft Center in Houston quickly designed and produced a zip gun—officially a *"handheld maneuvering unit"*—that fired spurts of pressurized oxygen. It should give Ed White some way to move here and there at the end of his umbilical tether. White practiced with it on a floating air bearing table and they added it to the manifest.

The space walk was the big news, but Gemini IV was more important than that. It was the first time the new Manned Spacecraft Center's mission control center was used. With nothing happening at the Cape after launch, reporters dashed for airplanes and flew to Houston. Some of them missed the lessons learned on Gemini IV's first orbit of Earth.

Quickly after reaching space, McDivitt turned the craft around and attempted to rendezvous with the Titan rocket's second stage. He hadn't been trained adequately and almost immediately discovered that an airplane pilot's reflexes didn't work in space. It wasn't all his fault; the orbital experts in Houston didn't understand some of the complexities either. McDivitt's frustrating experience taught all of them lessons.

If the rendezvous worked, Ed White would have approached the rocket stage on his space walk and photographed it. But the first thing McDivitt saw was that the stage was somewhere between 150 and 300 feet away. He did what a pilot would do: He pointed toward the stage and fired his thrusters. Amazingly it got farther away and seemed to drop below them. He fired the thrusters again and it happened again.

McDivitt and White looked at each other. McDivitt tilted Gemini's nose toward the rocket stage, lined it up carefully, and pushed his control stick forward again. The thrusters fired on order. Almost immediately the stage dropped farther below, and now it was moving ahead of them too. This wasn't right. They were discovering what the Russians had learned on their two tandem flights: Two ships in different orbits moved away from each other. They hadn't figured out that every time McDivitt fired his thrusters, he was moving Gemini into a slightly higher orbit. The higher the craft went, the slower it moved in relation to the rocket stage. Before they'd circled the world once, McDivitt had burned up half of Gemini IV's maneuvering fuel and only succeeded in widening the distance to the target.

"It's taken a little more fuel than we'd anticipated," McDivitt radioed as they approached the Pacific coast of Mexico. "Do you want me to really make a major effort to close with this thing, or to save the fuel?"

In Houston, Capcom Gus Grissom looked questioningly at the man called Flight. Chris Kraft didn't bother asking for other opinions. He made his first major decision in the new control center. "Save the fuel," he told Grissom.

That put Ed White front and center. He had an hour to get ready for his EVA—extravehicular activity—or space walk. It wasn't easy. The Gemini cockpit was considerably more spacious than Mercury's, but there were two men taking up room in their bulky space suits. They had to unpack the bag carrying the twenty-five-foot umbilical tube that was White's lifeline; it contained both oxygen and communications lines, eliminating any need for a Leonov-like backpack.

Getting the bag out and grappling with the snaky umbilical was more work in zero gravity than it had been on the ground. One end had to be plugged into a spacecraft receptacle. The other end went into White's suit. Both ends had to be checked and verified. This was no time to be sloppy. Suddenly they were across the Pacific Ocean and coming up again on Hawaii. There were still the cameras to get ready—a 70-mm Hasselblad for White to take still photos outside, a Hasselblad and a 16-mm movie camera for McDivitt to record the action from his left-seat vantage point. Nor were their gloves slipped on and locked. The setup work needed nimble fingers. Space gloves are not nimble.

"I don't think we're going to make it," McDivitt said. They kept working, but it was a no-go. "Next pass around."

In Houston, Chris Kraft nodded. "We're happy with that." Gemini IV passed over Australia for the third time in four hours, and finally everything was ready. McDivitt and White sealed their suits, flipped faceplates down and locked, and felt pressurized oxygen flowing around their bodies. White had the long umbilical bunched at his knees and feet, his hand ready to ratchet open the hatch. It looked good to the ground.

"Tell him we're ready to have him get out when he is," the man called Flight said to Grissom.

"Go for decompression and go for EVA," Grissom said, passing the word up immediately.

McDivitt flipped the switches to open valves and empty the cabin of oxygen. In quick minutes, they had the same vacuum inside as outside. White worked the ratchet . . . and nothing happened. The hatch wouldn't open. He pushed again, then again. Suddenly it popped free and he let out a long sigh. He pushed hard and the hatch opened wide. Near Hawaii, Ed White gave a little kick, floated up, and was free.

It was supposed to be a twelve-minute space walk. White tried out the zip gun, found that he could move this way and that with precision, then ran out of compressed oxygen just when it was feeling natural. "I've exhausted the fuel now and I was able to maneuver myself down to the bottom of the spacecraft and I was right up on top of the adapter," he reported. He sounded exhilarated. He tugged on his umbilical, moved and rolled easily in any direction he desired.

In Houston, Chris Kraft faced another decision. White's programmed time was running out, and Gemini IV was heading toward a communications gap between Hawaii and California. If he let White stay out there, there'd be several minutes without monitoring data. But the free-floating astronaut's data looked good. Kraft held off on the order to terminate the space walk.

Now McDivitt fired Gemini's small thrusters as White watched and said they looked good, no smoke or particles, just flame. Over California, then on to Texas, White maneuvered at the end of his umbilical while McDivitt did the photography that would become instantly famous a few days later. They were crossing familiar ground—"We're looking right down on Houston," White said excitedly—when Kraft decided that it was time to end the exercise.

"Tell 'em to get back in, Capcom," he ordered.

Grissom tried, but McDivitt and White were chattering to each other. The message didn't get through. White was having fun and McDivitt too.

"Gemini IV, Houston," Grissom tried again. "Gemini IV, Houston."

Finally McDivitt caught on. "Gus, this is Jim. Got any message for us?"

Chris Kraft flipped the switch on his console that let him talk directly to the spacecraft, one of the few times that would ever happen. "Yes," he said, his voice somewhere between stern and understanding, "tell him to get back in!"

"They want you to come back in now," McDivitt told White.

"This is fun," White responded.

"Well, get back in."

"I'm coming." They passed over Florida, and White was reluctantly lowering himself into the cabin: "It's the saddest moment of my life."

McDivitt knew the feeling, if not quite exactly. "You're going to find it sadder when we have to come down from this whole thing," he said.

Getting in was easy. Closing the balky hatch wasn't. They were nearly on the other side of the world before they had it latched down. When he was comfortable again, White reported that man could indeed control himself on a space walk. He'd walked out there twenty minutes and he'd outdone Aleksei Leonov in every way. Later he revealed that he'd taken a gold cross, a St. Christopher medal, and a Star of David with him to orbit. They represented his faith in God and "I think this is probably the most important thing that I had going for me on the flight . . ."

Gemini IV stayed in orbit four full days and did an extensive series of maneuvers to prove the ship's capabilities. Their ship was the first to have an onboard computer, a small unit by IBM built into the instrument panel. There had been worries about how it would work in space, even whether it would turn on. It worked fine on days one, two, and three. IBM executives, who had been sweating in the cold light of press coverage, felt so good that they ordered up a full-page ad in the *Wall Street Journal* extolling their minimachine. On that day in space, it wouldn't turn on.

John Young was the Capcom that shift, and the IBM brass winced when they heard him offer his suggestion to McDivitt and White: "Why don't you just kick it?" The computer stayed off for the rest of the flight. But on future flights, it worked as advertised in the *WSJ.*

In the odd moments when there was nothing to do but look out the window in awed amazement, McDivitt and White thought about the argument over music. Comic Bill Dana and record producer Mickey Kapp by now were old hands and close friends with the astronauts. Astronauts stayed at NASA East, Kapp's apartment in New York, or at NASA West, Dana's home in the Hollywood Hills above Los Angeles, when they needed to get away from autograph seekers or just wanted a few days' rest. The visits were never publicized, never promoted. In L.A., astronauts knew where to find the keys to Dana's boat, in a slip at Marina del Rey, and would party there, or take it out for a short cruise.

When Kapp suggested that McDivitt and White take some music along, he got a cold shoulder from the flight planners in Houston. The onboard tape machine was for the astronauts to record their observations or make private comments that might not be appropriate over the radios.

"It uses tape cassettes, right?" Kapp asked.

"Right."

"So how about if I provide cassettes with music?" he responded. "The guys can record right over it. But until they do, they'll have music with dinner."

It was almost too logical. The planners still said no, even with McDivitt and White siding with Kapp. So the only sounds the Gemini IV astronauts had were the static in their headsets and the clicking and clanking of Gemini motors and valves behind their seats. When Kapp tried again on Gemini V, the answer was the same. So he collected a case of tape cuttings and sent it to Gordon Cooper and Pete Conrad. The accompanying instructions called it a Do-It-Yourself Music Kit and explained how to splice tape together. After that, the recalcitrant equipment managers got the message and Kapp provided Gemini astronauts with music, but only broadcast from the ground.

Jim McDivitt and Ed White got their music from marching bands in parades when they came home heroes. They'd more than doubled America's longest previous spaceflight, and the next Gemini mission would double it again. White's space walk inflamed American imaginations and put the space race back on the front pages. For the first time, it looked like the United States might have a chance to win.

Gemini V was a turning point. The news media poured into Houston for the eight-day Gemini V mission in August 1965. Gordon Cooper and Pete Conrad were about to eclipse the Russians for stay-time in orbit. The Associated Press sent twenty-five people. United Press International sent eighteen. Nobody had a good count on the networks, with their anchors, reporters, technicians, and researchers.

Reporters clogged the streets in front of the homes of Cooper and Conrad, hoping for a look at the kids or a quote from a wife. They were there for another reason too. They called it the deathwatch. Death wasn't likely, but a news media deathwatch was a space mission must. *You never know*, editors said when they made the assignments. An open field near Conrad's house sprouted a sixty-foot microwave tower, flanked by an aluminum trailer, three phone booths, and a single Porta Potti. One reporter counted thirty-five cars parked in the street, including her own. The downside of an open space program was painfully obvious to the neighbors. But even they got into the excitement, bringing lemonade and cookies to reporters sweltering in the August heat and humidity.

Bob Gilruth's power and influence over America's manned space program was unassailable. But he began to move into the background, preferring to let the bright lights of publicity fall on some of the key people he'd nurtured and watched grow. His dry humor stayed intact. One morning Paul Haney had a small dental bridge placed in his mouth. That afternoon he, Gilruth, and a couple of Manned Spacecraft Center people were on a NASA plane to Washington. Gilruth had his usual martini before they broke out the sandwiches, and while they were munching, Haney's bridge came out. He swallowed it. Gilruth looked at the others on the plane and shrugged. "Well," he droned, "Paul will just have to pass that bridge when he comes to it."

Time chose Chris Kraft for its cover leading up to Gemini V, and its story treated him like a king. Kraft was just forty-one and he was the old man already. The average age of the 568 people in his Flight Control Division was just thirty. On many missions, that average dropped to twenty-seven or twenty-eight for the men inside mission control. Kraft now had three teams of flight controllers to work eight-hour shifts during long missions. The other two were headed by Gene Kranz, thirty-two, and John Hodge, thirty-four. Kraft liked to have the young guys in mission control. "They're more eager and they get caught up in what they're doing," Kraft said. "At Langley everyone left at 4:30. You won't find that around here. And sometimes we have to chase them out of here and tell them to go home." Kraft would still be there to do the chasing. He earned his $23,000 salary.

He'd already earned his reputation for making quick decisions. "The conductor can't play all the instruments," he told a *Time* correspondent. "He may not even be able to play one. But . . . he mixes all this up in a pot and out comes music. That's what we do here."

Kraft's prespace life as a topnotch aviation engineer and flight researcher was never far from his mind. *Time*'s Houston bureau chief, Ben Cate, joined Kraft at the Cape while researching the cover story, and they flew home on National Airlines. The flight hit severe thunderstorms on the approach to Hobby Airport in Houston. Kraft stared intently out the window as the Boeing 707 let down through the storm. Lightning flashed, rain scrolled across the windows, and the turbulence rocked the jet. Finally Cate leaned over: "Can you see the ground, Chris?" The answer left Cate with watery bowels.

"The ground?" Kraft snarled. "Hell, I'm not worried about the ground. I'm watching the wings. They can fall off, you know!"

Sometimes accidents happened. Kraft flew to the Cape just before Gemini V for one of the final prelaunch meetings. Waiting to go home at the so-called skid strip after the meeting, he watched helplessly as a limo driver backed over his suitcase. The suitcase, a shirt, a tie, and an unopened bottle of Cutty Sark were destroyed. Kraft submitted a claim for $61.30, along with a letter from Paul Haney, who saw it happen. NASA's bean counters disallowed the Scotch and paid him $56.05.

For a supreme techno-boss, Kraft had some superstitions. He wore a Mercury lapel pin during missions, but never otherwise. That pin became a constant during Gemini downtimes, to be replaced during missions by a Gemini pin. He would do the same kind of thing in Apollo. At the safe end of each mission, Kraft indulged another superstition, his good-luck cigar. He'd lean back, put his feet on his console, and light up. Before long, most of his control center team was doing the same thing.

Kraft got along with most reporters, especially those who got smarter and more sophisticated as the space race ran on. More than six hundred signed in for Gemini V, and on later missions, that number was dwarfed. Walter Cronkite helped solidify his reputation as the most trusted man in America during his live television coverage. A newsmagazine correspondent quoted another reporter in a dispatch to New York: "You have to do your homework to cover these flights," the reporter said. "The whole space field is a unique refinement of every other science."

NASA officials began to enjoy press conferences because reporters learned to ask good questions. Or most of them did. Bill Hines was a veteran space reporter for the *Washington Star,* but he could be belligerent. At one press conference, he was a bit too demanding. Paul Haney was running the show that day and fixed Hines with a bleak stare: "Would you care to snarl that question over again, Bill?"

Chris Kraft took pains to give clear and accurate answers. But when a new guy from the *Beaumont Enterprise* asked a particularly stupid question during a Gemini V press briefing, Kraft leaned into his microphone with a sigh: "Let me give you the facts instead of answering your question."

At another briefing, one reporter kept insisting that the astronauts were doing secret military experiments up there. "Yeah," Kraft responded, "we read some of your stuff like the funny papers." The reporter shut up. But the press had their own gotcha ready the next day. When Kraft sat down to begin a briefing, reporters held up newspaper comic pages. One of them gave him a fresh copy of *Time* with cartoons and gremlins in place of the regular cover artwork. He took it in laughing good humor.

Bob Gilruth was nervous about packs of reporters roaming the space grounds during missions. To handle the hundreds and eventually thousands of reporters who crowded into Clear Lake for each manned mission, NASA rented a two-story office building just across NASA Road 1. It put in twelve hundred phone lines for the press, in addition to its own phone. One area was turned into a four-hundred-seat briefing room, with press conferences held three times a day.

Live audio and video were brought in from mission control. A public affairs officer in the control center provided running commentary, along with live conversations between the astronauts and the Capcom. Within hours, typed transcripts of those sessions were handed out to reporters, though the typing pool often misunderstood words uttered in the special language of spaceflight. Reporters did their own translations.

NASA's Washington chief of public affairs, Julian Scheer, was generally satisfied with the news media. "There is a new breed of reporter who has to be a scientist and engineer and who has to have the instincts and speed of a police reporter," he said. "This man has to explain incredibly intricate problems against the pressure of a deadline. The exasperation comes when *we* can't present the explanation fast enough."

Gemini V set the stage. Gordo Cooper and Pete Conrad didn't die during eight days in space. They flew for the first time with a sophisticated rendezvous radar and did all the right maneuvers to show that rendezvous worked. Their prime electrical source was a set of fuel cells, which converted hydrogen and oxygen into water and electricity. The cells were balky, but they worked. Next up was Gemini VI. That was the one that might settle the question everyone was asking:

Who's really ahead in the space race?

The big complaint by reporters was their exclusion from mission control. Writers could watch the training and practice simulations in the weeks leading up to a mission, and a few did. But no reporter was allowed inside during the real thing.

The few who watched the last big training session in mission control before Gemini V saw a classic. It lasted six hours, with Gordo Cooper and Pete Conrad reading scripts and flipping switches in the Gemini trainer and a flight control team headed by John Hodge sweating at their consoles.

The scriptwriters were called devil's advocates. They wracked their brains to come up with problems that the controllers then had to handle. On this simulation, they had Conrad getting sick almost immediately after reaching orbit. The medics in the back room and on the surgeon's console kept close tabs until Conrad said he was feeling better. But almost immediately there was trouble with the radar pod to be ejected and then used for a practice rendezvous. (On the real Gemini V, the pod and the rendezvous practice all worked correctly.)

"We can't tell for sure, but we think it didn't eject," Cooper reported. The experts in mission control huddled, and Cooper turned the spacecraft around to look for the pod. It wasn't visible, he said, reading his script. After thirty minutes, the experts couldn't figure an alternate way to eject the pod. Flight director Hodge put his team to work designing an alternate flight plan, with maneuvers they'd monitor by radar from the ground.

"Why not?" he asked rhetorically. "We've got plenty of fuel."

What they didn't have was time. "We're getting some low O_2 readings," Cooper reported. The devil's advocates who wrote this training script had packed it with trouble.

After a quick analysis, the environmental guy on the EECOM console verified a leak. The eight-day flight might have to end after two days. Once more Hodge put his men to work revising the flight plan. "Move up as many experiments as possible," he ordered. "Let's keep the astronauts busy."

Now the oxygen leak got worse. "We're going to run out of primary O_2 at GET 18:00," EECOM reported. GET was ground elapsed time, the amount of time since liftoff. All flight plan events on American manned spaceflights are timed according to GET.

The secondary oxygen system had a twenty-four-hour supply. Hodge and his team decided that Gemini V would reenter after its eighteenth orbit. The splashdown point was designated 18-1. Then Conrad got sick again. He consulted with the medics, following his script exactly and complaining of nausea and stomach pains.

"He's no worse than he was," Surgeon told Flight. "I'll let him go, but he has a definite medical problem."

Hodge asked Recovery to look at bringing them down early. But each of the alternate splashdown sites, designated by numbers 2, 3, or 4, involved long waits for a recovery ship. "We can get to him faster if we use 18-1," Recovery said.

At GET 4:01, Conrad reported more cramping. Seven minutes later he talked with a Surgeon at the Carnarvon, Australia, tracking site. "He has cramps and tenderness," Carnarvon told Houston. "We're saying a possible diagnosis of appendicitis."

"That's our feeling too," said Houston Surgeon.

A few minutes later Conrad was on the radio again. "Cramping pains are getting pretty severe and I'm feeling nauseous," he groaned realistically.

"Take Item F," Houston Surgeon ordered. Item F was an antinausea pill in the onboard medical kit. Long minutes passed before Surgeon officially called it as appendicitis and said that Conrad should come down as soon as possible. He could wait twenty-one hours for recovery area 18-1.

"I can have an aircraft with jumpers on the scene for a landing at 6-4," Recovery said. That was still five hours in the future.

"O_2 leakage is increasing," EECOM interrupted.

"The pilot just vomited," Cooper radioed urgently.

"The pain is spreading across my abdomen," Conrad added.

John Hodge's flight control team was getting tense. They'd never faced anything like this, and it sounded very real. "We've got to get medical to him ASAP," Surgeon said.

"If they come down to 5-4, they'll be on the water for ninety minutes," Recovery interjected. "It'd be better to be in orbit and come in on 6-4."

Cooper wasn't happy with that. The scripted Conrad was still vomiting and now he had diarrhea. "The knobs are getting slippery," Cooper read from his script. Controllers in Houston drew graphic visual images of conditions inside the weightless spacecraft.

"I'm through," Conrad read from his script. "I want to come back."

"Give me 4-4 reentry figures just in case," Cooper demanded. "It's getting to be a mess up here." Recovery interrupted with a new estimate that a rescue plane could be waiting at 5-4 only ninety minutes later. Cooper didn't like that idea either. "Pilot is no-go," he said, almost gasping. "I think we better come down."

The Surgeon got on-line to Conrad. "Did you keep the pills down?"

"I took them. I think they're still there."

"Any more diarrhea?"

"I lost count."

"How about your suit?"

"I can't close it."

"Why not?"

"I don't dare."

Cooper pushed again for an immediate reentry into area 4-4.

"You'll be less comfortable on the water," Hodge told him, opening a direct communications line.

"Yeah, but at least everything will settle to the bottom," Cooper said. It sounded like he was talking through gritted teeth. "It's making me pretty sick too."

Hodge talked to his team. "It's a weird situation . . . he wants 4-4. Under the circumstances, it's their prerogative and we'll let them come in." Recovery area 4-4 was in the mid-Pacific, well north of the Philippine Islands. A Navy destroyer was dispatched in that direction, along with rescue aircraft.

"We're coming in on 4-4," Cooper radioed. Now he sounded determined. Minutes later mission control got a new message from Australia:

"He made an OAMS burn and is on the way down."

That shocked John Hodge. OAMS was the orbital attitude and maneuvering system, the big rockets used for rendezvous maneuvers. They were not meant for reentry. But apparently Cooper had fired them anyway, then jettisoned the adapter section and fired the retrorockets too.

"Where the hell did he get that?" Retro asked.

"He really is an independent son of a gun," Hodge muttered.

Recovery did some quick calculations: "If he really did it, it looks like he'll land a little short."

Hodge was almost afraid to ask. "How short?"

"About nine hundred nautical miles."

There was nothing there—no ships, no aircraft, no islands. There was quick conference in mission control.

"Call them quick," Hodge ordered. "Tell him to use a full lifting reentry so he can pick up another hundred and fifty miles or so."

Recovery was now getting radar reports of Gemini V's actual downward trajectory and began scanning Navy reports looking for a ship, any ship. He found one. "Flight, Recovery . . . The new impact point is very close to a ship, Flight. We'll get them."

It was the first moment of relief in mission control for six hours. Hodge sighed. "Okay, it's over . . . it's over." It would start again on Gemini VI, and for as long as Americans fly in space, training sessions, not always as gritty as that, would continue.

"The country that does the first rendezvous and docking will be ahead. No question about it."

Wally Schirra leaned back in a government-issue gray chair in a back office at Building 1, the public affairs building at the Manned Spacecraft Center in Houston, and looked at his partner for confirmation. "No question about it," said Tom Stafford. "You've got to have rendezvous and docking to do Apollo."

"We're going to do it and we're going to be first." Schirra grinned. "Write that down."

Schirra and Stafford were six weeks from the planned launch of Gemini VI in late October, doing the obligatory round of private press interviews and feeling good. Their mission, they were convinced, was the one that would put the United States unequivocally ahead of Russia in the race to the moon. No Russian had flown since the spacewalking Voskhod mission the previous March. Intelligence reports were showing no sign of another mission anytime soon. For the Russians, that was unusual. They had a habit of pulling some space rabbit from a hat just before a major American flight.

Gemini VI was to be the first double launch in the American program. An Agena target vehicle would go first, atop an Atlas rocket. Agenas were developed as an upper stage for military satellites, and their single rocket engine could be started and stopped several times in space. For Gemini, the Agena was modified to carry a radar unit that the manned spacecraft could lock onto and home in on. The end where a satellite would normally sit was fitted with a docking collar and electrical connectors. If the rendezvous worked, Schirra would plug Gemini's nose into the collar and be locked tight. The electrical connectors would give his copilot, Stafford, control to fire the Agena's engine, though that was not planned on this first Gemini-Agena mission.

Schirra was a logical choice for the mission. His pilot's touch was legendary and he'd flown that textbook Mercury flight. He had the feel, and now, after the lessons learned in Geminis IV and V, he had the training. In a full-scale press conference the next day, he put it into perspective. "The art of rendezvous is quite complicated," he said. "We're . . . counting down four separate systems . . . the Agena, the Atlas, the Gemini, and the Titan. So they all have to fall together.

"Now *we* will launch, if we have a good Agena."

It turned out to be a big *if*.

Schirra and Stafford were buttoned up in their spacecraft on the pad at the Cape on the morning of October 25 when the Atlas-Agena completed a flawless countdown and roared away. Their own countdown was perfect too. They expected to be on their way in an hour, forty-one minutes. Six hours later, if Schirra was as good as everyone thought, they'd pull up behind the Agena about 161 miles above Earth and ask for Chris Kraft's permission to dock.

As the astronauts relaxed, telemetry data from the Agena made controllers sit upright and stare. It had separated on time from the Atlas, but it was wobbling. As its automatic stabilization system tried to steady it, the Agena main engine fired up to send it into orbit. Then the data flow stopped. Moments later radar was weakly tracking five pieces of falling trash instead of one middle-size rocket stage.

At first Schirra and Stafford misunderstood the garbled talk on the voice loop and thought all was well. But as the minutes dragged on and ground stations reported no luck in tracking the Agena in orbit, the truth soaked in. Schirra's colorful language left ears burning, but nobody faulted him for his disappointment. Rumors leaked out that Schirra's comments were less than diplomatic, and three days later at a press conference, a reporter brought it up: "Tom, could I ask you precisely what Wally said when you both learned that the Agena had blown up?"

Stafford glanced sideways at Schirra, put on a big grin, and answered, "Something like 'Gosh darn, it's been a bad day.'"

The worst of it was that getting a new Agena ready for them would shift their mission well into the first quarter of 1966. But the day hadn't stayed bad for long. The astronauts immediately flew home to Houston, and before they could find time to get a drink, they'd been called into a meeting. Gemini program manager Chuck Mathews put it to them:

"What would you think about rendezvousing with Gemini VII?"

"I had about four hours to get disappointed," Schirra told reporters later. "All of a sudden we're all elated."

Gemini VII was scheduled for early January, a fourteen-day spaceflight by two more of the Second Nine, Frank Borman and Jim Lovell. Within a few more hours, the new mission was being called Gemini VII/VI-A. The press ignored the Roman numerals. The mission became Gemini Seventy-Six, or Gemini 7/6. Rumors quickly circulated that it could fly in December. The American space team was getting good. The official date was early January. In Houston and at the Cape, everyone knew that they were aiming for early December, only six weeks away.

At 2:30 p.m., December 4, 1965, Gemini VII was on its way. Borman and Lovell had compressed their training for their long-duration test, launch pad crews had removed the Titan from Gemini VI, replaced it with the slightly more powerful rocket needed to put the heavier fourteen-day spacecraft into orbit, and were ready for the fastest pad turnaround anyone had imagined. With Seven on its way, they'd be ready to launch Six in only nine days. Frank Borman and Jim Lovell settled in for the longest Gemini mission planned. An extreme Apollo moon mission could take fourteen days, though all were expected to be less. But the data about human endurance and survival from the long Gemini mission were essential to planners; they had to know what impact fourteen days of weightlessness had on the human body.

It had plenty. When they returned, Borman and Lovell were shaky on their feet and Borman suffered from dizzy spells. Their muscles had weakened and atrophied—in part because they were essentially sitting in the front seat of a Volkswagen for two weeks and couldn't do any real exercise—and blood pooled in their lower legs. For parts of their flight, they'd been able to take off their space suits and get more comfortable. Sitting in a shirtsleeve environment made life more bearable. Within a day, both felt almost normal. The flight surgeons, even those who had expressed the greatest fears about man-in-space before Mercury, were happily convinced. Man could survive long periods in space. Man could go to the moon.

Wally Schirra and Tom Stafford strengthened the case. Their Gemini-Titan was moved to the launch one day after the Gemini VII launch. Checkouts and countdown consumed the next week. They were suited up, sealed in, and ready to go on December 12. There was a hold at T minus six minutes to let the launch pad team get the timing exactly right. Six was to climb into orbit just behind Seven. At 9:54:06 a.m., the Titan rocket roared. At 9:54:07.2, it went silent.

A cloud of red smoke—the normal result of the Titan's propellants burning explosively—roiled up from the launch pad. Nothing else happened. In the launch control blockhouse, Bill Dana blanched. He'd been brought into the restricted building by Gus Grissom, who was cussing quietly while they waited for . . . what? The next seconds could mean life or death for the Gemini VI crew. Schirra and Stafford sat frozen in their spacecraft seats while the Titan shivered and went still. Schirra thought about grabbing the ejection seat D-ring—he was supposed to be clutching it at the end of the countdown—but decided not to. One quick pull and they'd be blown outward and upward, theoretically hurled safely away from an exploding rocket. He knew the risks; a Gemini ejection was likely to cause some injuries. He knew the other certain result: pull that ring and this mission was over. It would be months before the spacecraft could be ready to fly again.

"I never touch it [the D-ring] anyway," Schirra told friends at a wind-down party that night. "If there was nothing wrong, I don't want to be sitting on the beach while the rocket takes off for orbit."

This time there was something wrong. Almost simultaneously, each astronaut realized that he was holding his breath. They let their lungs exhaust slowly, listening for any sound that might be the beginning of an explosion. There was silence. Again slowly, they felt the tension in shoulder muscles relax. They looked at each other and Schirra nodded. *Okay, we're okay.* He glanced at the untouched D-ring. It was safe to breathe again.

Those few moments of icy calm saved the mission. The rule was written large and Schirra knew it. He should have yanked that D-ring. But it didn't feel right. It felt . . . safe. The man's feelings confronted the machine's rules and the man did the right thing. Technicians found that an electrical plug had come loose a second or so too soon. The computer read the loss of power and shut down the engines. There was more. When they inspected the engines later that day, they found a plastic cap lodged inside a propellant line. It had been there since the Titan left its assembly plant. Schirra and Stafford understood the implication: The rocket would have shut down back in October too. Getting this one off the ground was no easy chore.

They did it three days later. Seven was up there, in a circular orbit 185 miles high. Six was coming. Seventy-Six was about to make some space history.

Shorty Powers was at the pre-mission press conference as a freelance writer. He'd been fired by Bob Gilruth when he'd gotten too aggressive in his job, almost taking on the aura of an astronaut himself, and stayed in the Air Force only long enough to retire. He'd pressed Schirra about the meaning of a rendezvous without docking. "The challenge is to get that relative velocity between the two vehicles down to one, or less, feet per second,"

Schirra answered. "When you get down to that eyeball-stick-and-rudder type of work, the last three or four miles is the real challenge." Nose-to-nose, Schirra said, was his definition of a rendezvous.

Jim Maloney of the *Houston Post* wasn't satisfied. Why was this mission any different from what the Russians had done with their Vostoks? Schirra snorted and was derisive. "They were passing acquaintances," he said, with almost nine thousand feet per second of maneuvers needed to make it a real rendezvous. Their maneuver capability was less than that, nine thousand feet less. Schirra intended to park in Seven's garage. Now it was finally time to get to work. Six went perfectly into a lower orbit, nearly 1,400 miles behind Seven. Lower meant they were catching up. By the time they got around to New Orleans, they cut the distance to 730 miles. It was time for Schirra to go to work and for all of the Gemini and Apollo people to find out if this rendezvous thing was really possible.

Pointing Six exactly forward, Schirra fired its orbital thrusters. He only added thirteen feet per second to their velocity, but it was enough to raise their apogee on the other side of the world from 161 miles to 169 miles. They were still catching up, but more slowly. The physics of space travel was working: add speed at perigee to raise your apogee and vice versa. So he did it again at apogee over Australia. Now the perigee went up from 100 to 139 miles. Next came a vital maneuver. They called in an "out-of-plane translation." Six and Seven were on slightly different orbital tracks. Schirra pointed Six's nose south and fired his jets again. The spacecraft nudged sideways onto the same track as Seven. They were only 300 miles apart.

An hour later the radars locked on. Six was homing on Seven like an airplane looking for a runway. They were 270 miles apart. Schirra and Stafford could now track Borman and Lovell on their instruments. Schirra added velocity again, raising Six's orbit to a near circle at 169 miles up. He'd added forty-three feet per second to the closing rate, and it was time to coast for a few minutes. Inside both spacecraft, astronauts were scanning the sky for any sign of each other. Schirra saw it first, a bright star ahead. "That must be Sirius," he reported. It wasn't. It was Seven, only 62 miles away.

"Terminal phase" was at hand. The radar told Schirra and Stafford that they were on the right track and catching up. Schirra fired his thrusters, waited, fired again, waited, fired again. The star ahead was getting brighter, and fast. Six was closing on Seven at better than sixty miles an hour. At a half mile out, Schirra started to brake by firing his forward thruster. Six coasted in. Schirra braked again, slowed, braked again, and brought Six to a relative dead stop 130 feet from Seven.

This was no passing acquaintance. The spacecraft faced each other nose-to-nose. It was rendezvous by anybody's definition, Seventy-Six together in orbit.

Far away in mission control, Chris Kraft turned from his flight director's console and broke tradition. Instead of waiting for mission's end, he took out two cigars, handed one to Bob Gilruth, and they both lit up. On the big television screens in the news center, reporters saw normally stoic flight controllers cheering and waving American flags that appeared from somewhere. More cigars lit up and a blue haze filled the control center. Newsmen and women were cheering too, those who weren't on the phone filing bulletins and new ledes to their newspapers or already at their typewriters pounding out copy they'd have to rush over to the Western Union office in the next fifteen minutes.

The rest was fun in orbit. Frank Borman and Jim Lovell had been up there for eleven days. The sight of another human face, even across the way in another spacecraft, was more than welcome. They were bearded, grungy, and smelly—"Imagine spending two weeks in a men's room," Lovell said later—and it made them feel almost human again.

Schirra's fine touch had used less fuel than expected. Six had plenty left to nuzzle up to Seven, to fly around it so that Stafford could take pictures from every angle. After the complexities of getting there, this part was easier for Schirra than flying formation in airplanes. He found that station-keeping in orbit meant that you could do pretty much whatever you wanted. At one point, he brought Six to within inches of Seven's nose and had no trouble with control. If they'd been equipped for spacewalking, it would have been easy for Stafford and Lovell to switch ships. Finally it was time to part. Schirra had one last message for Frank Borman, who graduated from West Point, then went into the Air Force. He held up a neatly lettered sign that brought a guffaw of laughter from the Seven craft: BEAT Army.

Schirra backed off to ten miles and the crews got a night's sleep. Before they nodded off, Mickey Kapp's music finally got into space. Kapp wrote fresh lyrics to "Hello, Dolly," Jack Jones did the singing, and at Paul Haney's intercession, the song was broadcast up from mission control. Schirra the Navy man and Stafford the Air Force pilot laughed out loud when they heard it.

Hello, Wally. This is Jack Jones, Wally.
It's nice to know you're up where you belong.
All systems go, Wally, you're 4–0, Wally.
Tom, that's Navy jazz for razz ma tazz,
you can't go wrong.
And while the Earth's turning,
midnight oil was burning,
to bring you your old favorite songs
from way back when,
So, set the wax fellas,
loosen up and relax fellas.
See you back in Houston town again.

"Set the wax" was the cue to mission control to roll tape on a long playlist of records for broadcast up to space until both crews had fallen asleep. The next day, Schirra and Stafford did their own holiday vaudeville act from orbit. Stafford excitedly reported spotting another object in orbit: "looks like a satellite going from north to south, probably in polar orbit." On the ground, controllers perked up in puzzlement and reporters stopped their chatter to stare at the television monitor (which only showed the control center) and to listen to the loudspeakers.

"Stand by one." Stafford's voice didn't sound worried. ". . . let me try to pick up that thing." Reporters looked at each other with furrowed brows. Flight controllers fiddled with their headsets and seemed to be adjusting volume knobs on their consoles. *What the heck . . . ?*

It took a second for the next sounds to register. It was a four-hole harmonica playing "Jingle Bells." And there were the bells too, joining in with a merry jingle from orbit. The newsroom erupted into laughter and song, and reporters could see that controllers across the street were joining in too. Mickey Kapp had slipped the harmonica to Schirra, and he'd smuggled it into orbit with Deke Slayton's knowledge. Stafford's little bells came from Fran Slaughter, a secretary at the Cape, and Slayton had approved them too. It was, after all, the week before Christmas. This time, no one in Congress complained about the contraband.

A few hours later Schirra pulled off the last important part of the Six mission, a manually controlled lifting reentry. Gemini's design gave it some aerodynamic lift, and by turning left, right, left again, Schirra could go long or land short of the point indicated on his instruments. A lifting reentry would be mandatory in Apollo, which would come back from the moon and plunge straight into the atmosphere. "If we can't prove the lifting reentry," Schirra had said, "kiss Apollo good-bye." He handled it perfectly. Six landed just eight miles from the aircraft carrier *Wasp,* and for the first time, a splashdown from space was carried on live, satellite-relayed television. Seven came home two days later, and Frank Borman won his bet: He splashed down on live TV a mile closer to the carrier than Wally Schirra had.

The year 1965 was the turning point in the space race. Bob Gilruth issued a statement calling it "a fabulous year for manned spaceflight." After that single Voskhod flight and space walk, no Russian had flown. The United States flew five Gemini missions in less than eight months, put ten astronauts into space, did a space walk and a rendezvous, and set endurance records of eight days, then fourteen days in orbit.

There was no sign of activity from the Russians, and that was curious. But then, nobody knew that the Chief Designer was dying.

His bowel cancer was killing him. The Chief Designer hadn't told anyone how sick he was, and when he finally went in for polyp surgery—the hideous cancer still undiscovered—he let even his closest associates think that it was a simple procedure for hemorrhoids. Sergei Korolev died on the operating table January 14, 1966. The long operation to remove polyps and some of the surprisingly large cancer was apparently successful. But his heart, more damaged than even he suspected, gave out. He was fifty-nine.

Korolev's death lifted the secrecy surrounding his life. The Soviet news media published full obituaries, chronicling his achievements and identifying at last the mysterious Chief Designer. His secret awards included the coveted Lenin Prize.

He was given a state funeral attended by the country's top leaders and lay in state in Red Square with an ever-changing honor guard of scientists, engineers, cosmonauts including Yuri Gagarin, and government leaders. Among the latter were Party Chairman Leonid Brezhnev and the KGB's Yuri Andropov, who stood their turn silently at guard. The nonperson Nikita Khrushchev was not allowed to attend.

Sergei Pavlovich Korolev's ashes were interred in a Kremlin wall, in a section reserved for the most honored of state heroes. A black marble tablet covers his urn.

For the Chief Designer, the race was over.

11

They called Neil Armstrong "the astronaut who thrives on danger." No one knew as he and Dave Scott, the first of the New 14 to get a mission, trained for Gemini VIII that the danger was just beginning.

Armstrong was an iceman at work and at home. His wife, Jan, a dark-haired beauty, had learned in their years together that he was contained within himself, a nondemonstrative personality who seldom let loose. She adjusted and adapted until reporters who tried to get her to smile privately called her the Ice Maiden. Jan Armstrong was just what Neil needed at home. Dave Scott, with a keen engineering background and reflexes like a cat, was what he needed at work.

Near-death experiences formed Neil Armstrong. He earned his Navy wings at nineteen and at twenty-one was flying sorties from the carrier *Essex*, jockeying a Panther on strikes into North Korea. He was a hot pilot, fearless in flight, with reflexes even quicker than his future crewmate, Dave Scott, and the ability to analyze events and make fast decisions. That saved his life twice in Korea. He was bridge-busting behind enemy lines in September 1951 and spotted a truck convoy. Armstrong swept in low, triggering cannon fire, getting a world of flak in return. Chunks of his Panther's wing and tail section tore away. Shards of metal ripped through control cables and severed hydraulic lines. He nursed the plane back to altitude, realized that its time in the air was short, and pointed its nose south. Somehow he kept it flying across one hundred miles of Chinese-controlled territory. He was just over U.S. lines when the jet engine choked and quit. Armstrong didn't hesitate. He ejected and came down under his parachute among friendly troops.

A few months later he saw enemy troops congested in a narrow valley. He couldn't see the trap. Armstrong swept up and over, came down fast and strafing into the valley, and was almost immediately tossed sideways when the Panther jolted hard and almost flipped over. He'd flown into a cable strung from valley wall to wall. It ripped away a wingtip, but the jet was flyable. He limped it back to the *Essex* and landed safely. "All the pilots had close calls," Armstrong shrugged when asked to tell the story. One of his favorites was getting clobbered by cannon fire and flak on this seventy-eighth and final mission. He got back again and walked around his Panther counting the holes. There were seventy-eight.

He mustered out, went to Purdue University for an aeronautical engineering degree, got married, and got bored. NACA hired him as an engineer at its Cleveland facility, then transferred him to be a test pilot at Edwards Air Force Base, California. He flew them all, the F-100, F-104, B-47, and F-102. Then he switched to the NASA rocket planes, logging flights in the X-1 and X-14. Walt Williams took him into the X-15 program, forgave him for oversleeping, and put him to work flying the B-52 mother ship. Finally he got his chance to fly the X-15, twice in 1960, once in 1961, four times in 1962. Each time was an event. One came close to being deadly.

He dropped away from the B-52 on a 1962 flight, flicked through the rocket plane's starting procedures, and heard only silence. The X-15 was dropping like a rock toward the desert floor. Armstrong gave it a half second's thought, then started over. The rocket ignited with a roar and again he didn't hesitate. He pointed the X-15's nose toward the sky and flew to the edge of space. When the rocket fuel was gone, Armstrong was flying close to Mach 6, 3,818 miles per hour. He kept the nose up and coasted to 207,000 feet, where the sky was black and Earth was a long, curved horizon in any direction he looked. He liked what he saw and felt. Five months later he was in Houston, one of the Second Nine group of astronauts.

For the next two years, he trained and did the grunt work to help other astronauts fly Mercury and Gemini missions. Life was good until the dark of a chilly spring morning in 1964. Jan was choking and coughing next to him in bed, and he woke to hear flames crackling through their new house in the village of El Lago, just three miles from the Manned Spacecraft Center. The two ran down the hall to rescue Rick, seven, and Mark, one, then stood outside and watched the fire gut the place.

Faulty wiring was to blame. Quick instincts by both Neil and Jan Armstrong had saved their lives and their children.

Now it was time to ride the rocket again, March of 1966, and the Gemini VIII mission was on the pad. It was also a time to grieve again. The Gemini IX prime crew, Elliott See and Charlie Bassett, had been killed February 28. The weather was lousy at St. Louis Lambert Field, and their T-38 jet clipped the roof of the building at McDonnell Aircraft where their spacecraft was being built. Bassett was the second astronaut of the New 14 to die. See was the first to die of the Second Nine. Neither would be the last of their groups to give the last full measure in pursuit of a space race victory.

The Gemini 7/6 flights put rendezvous in the plus column for America. Armstrong and Scott intended to add the last bit needed to prove that the Apollo plan would really work. They'd rendezvous with an Agena target vehicle, then dock with it to create a single joined ship in space. Once checkout was completed, Scott was trained for a space walk. He'd deploy a micrometeorite detector, try several kinds of space tools stored back in the Gemini adapter section, and try moving around without an Ed White–style zip gun. But first he and Armstrong had to get into orbit and do the docking.

It started out perfectly, an Atlas rocket putting the Agena target vehicle into orbit on March 16, then Gemini VIII following about ninety minutes later. It was the first manned space mission without Chris Kraft as a flight director. He'd moved on to planning Apollo operations; the first manned Apollo flight was only a year away. Two flight control teams, headed by John Hodge and Gene Kranz, were ready to take over. They planned to work twelve-hour shifts during Gemini VIII. New teams headed by Cliff Charlesworth and Glynn Lunney were in training and would be ready for later Gemini missions. Kraft was technically out of the loop, but all the teams ultimately worked for him and he had a habit of showing up in mission control whenever anything exciting was happening. At some of those times, he'd prove his trust in his people by gritting his teeth and not interfering. Gemini VIII was about to give him the first of those times.

The little problems started early. With each rendezvous maneuver, Armstrong saw that the thrusters didn't shut down cleanly. There was some small residual thrust. After the third one it looked better, but suddenly ground control ordered an additional firing to correct a velocity error. Then everything seemed to clear up. For the next three hours, Armstrong and Scott closed on the Agena rocket, spotting it first as a shining star about eighty-seven miles ahead. The rest was easy. Less than six hours after launch, Armstrong braked to a stop about fifty yards from the target. If there'd been any doubt about rendezvous techniques, they were permanently erased. Armstrong flew patterns around the Agena while Scott inspected it, making certain that nothing looked wrong. They were go for docking.

Now Armstrong showed just how keenly an astronaut could control a spacecraft in zero gravity. He lined up with the Agena's docking adapter and used the small Gemini thrusters to crawl forward at just three inches per second. The closure rate was so slow that it took the human eyeballs of Armstrong and Scott to know what was happening. Then it was done. "Flight, we are docked!" Armstrong called, and there was emotion and happiness in his voice, replacing the crisp test pilot patter that the control center expected from him. "It's a smoothie."

It took a moment for it to sink in back in Houston. Then both the newsroom packed with reporters and John Hodge's control center team erupted with cheers. The last theoretical barrier to going to the moon had just been removed. One spacecraft had done a precision, on-time rendezvous and docking with another in orbit. While the celebration roared on, the astronauts went back to work. They were looking right into the Agena control panel, only a few feet from their windows. The first task was for Scott to take control of the Agena and use its thrusters to turn the whole docked set sideways in orbit. They were now out of radio contact with Earth, over the Indian Ocean and between ground stations. The last message they'd heard from Capcom Jim Lovell was ringing in their ears.

Data signals from the Agena hadn't been clear, and no one on the ground was quite sure that it was acting right. "If you run into trouble and the attitude control system in the Agena goes wild," Lovell had radioed, "turn it off and take control with the spacecraft." Dave Scott's instrument panel gave him control of the Agena from inside Gemini VIII. Even without Lovell's warning, he was ready for emergencies. But the sideways maneuver seemed to go just right. Then hell came aboard.

"Neil, we're in a bank," Scott said calmly. His instruments showed that Gemini was rolling over. Armstrong's needles agreed. He blipped the thrusters and the roll stopped. When he relaxed on his hand controller, it started again. Armstrong fired again, bringing the motion to a dead stop. Now the astronauts were puzzled. *It has to be the Agena.* Armstrong relaxed and the docked Gemini/Agena took off again, rolling over more quickly. *This is getting out of control.*

The weak point was at the docking collar where the two craft were locked together. While Armstrong struggled to get control, Scott grabbed a camera to get photo documentation of the stresses. Then he used his own control switches to turn off the Agena completely. The roll continued; the off-command was being ignored, he thought. He tried again. Nothing. The roll rate was increasing and getting uncomfortable.

"We were jammed in there shoulder-to-shoulder," Armstrong said later, "and pretty much sitting over the center of gravity. That one fact made it easier on us." It wasn't like they were being whipped at the end of a string. Not yet.

They were still out of touch with ground stations, but it was time to make a decision. Armstrong's gauges showed that 70 percent of Gemini's fuel was already gone and the needle was dropping. "Time to undock," Armstrong grunted.

"Roger that," Scott replied.

Armstrong made one last effort with the thrusters to slow the roll rate. It worked and he said one word: "Go!"

Scott's finger was already on the button that would release Gemini from the Agena. He punched it and Armstrong backed them away quickly. In an instant, they were rolling so fast that their heads bumped. A pitch movement was added to the roll. At that moment, they passed above the horizon, and the tracking ship Coastal Sentry Quebec got them on the radio. "We have a serious problem here," Scott called. ". . . tumbling . . . disengaged from the Agena."

The words "serious problem" echoed back to Houston. In the control center and in the newsroom, everything stopped. There was silence as ears strained. "We're rolling . . . can't turn anything off." The voice was Armstrong's. He was cool, no panic inflections in his report. ". . . increasing in a left roll."

The roll rate was accelerating, already up to a full sideways spin every three seconds. It was getting worse. "We have a violent left roll here," Scott reported. He wasn't panicked either, just passing along the information like a test pilot was trained to do.

Armstrong's hand controller went dead. Nothing worked. The roll rate increased to a full 360-degree spin every two seconds. "I was getting gray around the edges," Armstrong said later, remembering how his vision was blurring. "It was getting uncomfortable."

There was one option left. Two rings of small thrusters, the reentry control system (RCS), surrounded Gemini's nose. It was a redundant set; one RCS ring had to be used for control during reentry. The other was a backup. They began punching circuit breakers to cut off the primary OAMS—orbital attitude and maneuver system—thrusters. By now both astronauts were getting dizzy and the roll was still getting worse. It had reached 296 degrees a second, almost a full spin every second.

Armstrong switched on the RCS and tried his hand controller. It worked. He turned off Ring B and grinned through his hazed vision. More by touch than sight, he fired the little RCS jets and started to slow the roll rate. Scott let out a big sigh. They were back in control and stable. It felt good.

Now they had to answer the big question: why? One by one, they reactivated the big OAMS thrusters. Number one. *Nothing.* Number two. *Nothing.* Number three . . . And finally when Scott punched on the circuit breaker for number eight, it all started again. They rolled violently to the left, and before it could do more than that, he punched the breaker off. Armstrong stabilized them again and now they knew—thruster number eight had failed in the on position. It had been firing continuously, burning up fuel, and damn near killing them.

There was nothing wrong with the Agena. It was their own Gemini spacecraft that betrayed them. While they'd stayed docked, the Agena was trying to keep them stable. When they'd undocked, hell came home until they solved the problem. The failure was electrical. A postflight engineering memo called the event a "unique chain of occurrences" which very effectively masked a simple problem. On future Geminis, a single switch would disable the entire thruster system; a violent spin could be stopped before it started. The reentry system wouldn't be needed.

But on Gemini VIII, the mission rule was clear, and back in Houston, Flight—now in the person of John Hodge—ordered them down. The RCS rings had been activated; there was no other choice. The best bet was a *"dash Three"* recovery area in the Pacific Ocean, on Gemini VIII's seventh orbit of Earth. They passed the information to the crew.

"I had no idea what they were talking about," Armstrong said much later. "What's a *dash Three?* We had to look it up in our onboard book." Then it was clear and the procedures were in writing too. They'd trained for alternate recovery areas. It was just a matter of setting up and doing it. No sweat.

The sweat, though, was flowing back home. Audio crews from the Manned Spacecraft Center installed squawk boxes, little black loudspeakers, in each astronaut's home during a mission so his family could listen to everything straight, not filtered by some television reporter's analysis. Moments after Dave Scott reported the problem, somebody in a back room at mission control turned off the boxes.

"I called the control center immediately," Jan Armstrong said, "to get permission to come over. They said no. So I went over to [the Scotts'] house. [Lurton Scott had] been in the control center when it happened and they ushered her out. They wouldn't let her stay."

The wives tried to find out more, but for a while they got their news the way the rest of the world was getting it—through television and radio reports. Within the hour, they knew that their men were safe and that Gemini VIII was coming down early. There was nothing to do but wait for splashdown.

"I was livid," Jan Armstrong said, revealing that she'd never been the Ice Maiden of reporters' assumptions. "But I had to hide my feelings. The excuse they gave us about not letting us in mission control was that they couldn't face us if something went wrong." Her tone of voice showed pure disdain for that line of thought. She made up her mind that if Neil ever got another flight, she'd set the rules about access to information, not be at the whim of some faceless person in the control center.

Jan Armstrong and Lurton Scott weren't the only angry people. The television series *Lost in Space* had been preempted by ABC to cover the real in-orbit drama. The ABC switchboard in New York City was clogged with callers protesting the decision and demanding that the network give them back their fictional entertainment.

Neil Armstrong and Dave Scott splashed down at dawn in the Pacific Ocean. They were in the middle of emptiness 450 miles southeast of Okinawa. But so precise was Armstrong's reentry, and so good was mission control's calculation of their position, that a rescue aircraft out of Naha was already there and saw Gemini VIII drifting under its parachute to the water. The astronauts didn't see the plane. They thought they were alone. "The only thing we could hear on the radio was Oriental music," Armstrong said.

Jumpers went into the water immediately and within forty-five minutes had a flotation collar attached to the spacecraft. It still wasn't a good boat. After their cool handling of the first American space emergency, the pitching and heaving of Pacific swells got to them. Especially the heaving.

It didn't go the way the Gemini people in Houston had hoped. But Neil Armstrong and Dave Scott had made space rendezvous seem routine and they'd successfully docked with the Agena. What happened next was another demonstration of man's value in space. But it didn't have much to do with what really mattered: using Gemini to prove the techniques that would take Apollo astronauts to the moon and back. By that measure, Gemini VIII did its job.

Gemini was over so fast that history sometimes forgets that it ever was. Yet it was a frantic, exciting, frustrating, and fulfilling time for the people who were there. Every two months or so through 1965 and 1966, another Gemini was on the way and astronauts were in orbit. In one way or another, every Gemini had its bit of trouble and every Gemini found a new way to test its crew and the controllers sweating it out in Houston.

From the beginning of Gemini manned missions to the end, no Russian flew in space. If there was still a race, only one country seemed to be running.

Gemini IX went into the books as the mission of the Angry Alligator. While Tom Stafford and Gene Cernan, the backup crew who became prime after Elliott See and Charlie Bassett died, waited on the pad on May 17, 1966, the Atlas rocket carrying their Agena target vehicle went astray. But data kept coming from the Agena for seven long minutes. When it was sorted out, they knew that both Atlas and Agena had gotten partway to space, then turned around and dumped into the Atlantic Ocean.

But this time there was a backup. After the Gemini VI Agena blew up, NASA had built a squat little replacement they called the ATDA—augmented target docking adapter. It was part Gemini adapter module, part docking assembly, all inert. It had no rocket, but it was something to aim at in a rendezvous and something to use for docking practice. They hauled it out of storage and launched it on another Atlas on June 1. Its orbit was perfect. Its condition wasn't. The cone-shaped clamshell shroud over the docking ring didn't fall off in orbit. When Stafford and Cernan did a quick three-orbit rendezvous, instead of the usual four, they were confronted with the strangest sight astronauts would ever see in space. The clamshell was cocked open like a set of swamp jaws.

"It looks like an angry alligator out there," he told Houston, and it did. At the Escape Velocity Press Club by the space center, an artist immediately drew a caricature of an open-jawed alligator. The club's board sent it out to a local jeweler, and when Stafford and Cernan came home, it presented them with Angry Alligator lapel pins.

They couldn't dock, but Stafford and Cernan could do a series of planned rendezvous exercises. They backed away and did two more of them that day and one the next. On their third day up there, Cernan did America's second space walk. This one was designed to test space suit mobility and to use a maneuvering backpack unit stored in the Gemini's adapter section. Just getting ready out there was more work than anyone thought. Cernan was floating behind Gemini, sweating, breathing hard, having trouble getting set to don the backpack without floating away. His faceplate fogged and he could barely see. Then it got worse and he couldn't see at all. There were space walk lessons there and they learned. One was to know when to quit. They told mission control that it was a no-go, and an exhausted Cernan climbed back inside.

Next up were John Young and Mike Collins, in July. Their Agena was waiting on orbit when they got there. Young docked and they fired up Agena's big engine to climb 474 miles above Earth, a new altitude record. Over the next few days, Collins did two space walks—one simply standing up in his seat and taking pictures, the other a free-float at the end of a fifty-foot umbilical.

Between the two, Gemini X proved out another aspect of rendezvous. After cutting loose from their own Agena, they caught up with the Agena rocket left in orbit by Neil Armstrong and Dave Scott. Collins used a zip gun to maneuver over to the rocket and picked up a micrometeorite collector that had been attached to it back in Florida. It was too difficult to attach a new one, though. He zipped back to Gemini X and they called it a mission. More lessons had been learned. One was that, even with new training techniques and improved equipment for maneuvering around, spacewalking could be tough.

Pete Conrad and Dick Gordon on Gemini XI in September did just what Apollo needed. Flight planning for moon missions now was aimed at astronauts in their lunar module doing a one-orbit rendezvous with their command module. Conrad and Gordon did it in Earth orbit, using a new series of maneuvers to catch their Agena just before they passed over the Cape for the first time. They docked, then both practiced docking and undocking. It was easier in space, Conrad said, than in the trainers in Houston.

There was a space walk again. Now they had a handrail along, and it made things easier for Dick Gordon. He straddled Gemini's nose like riding a bronco and, with more grunting and exertion than they figured, managed to attach a long tether connecting Gemini and Agena. The plan was to undock later in the mission and test the theory that two orbiting objects connected by a tether would stabilize themselves in position, without firing any thrusters. But first they fired up the Agena and set another altitude record, 853 miles above Earth. The ebullient Conrad could hardly contain himself. "Whoop-de-doo!" he hollered. The view was terrific.

They took it back down a few orbits later and went to work on another stand-up space excursion. This time Gordon spent nearly three hours doing scientific photography, standing on his seat and capturing images of star fields. Going outside in space was becoming routine. But even a stand-up was tiring. For a few minutes, they relaxed and both of them fell asleep— Pete Conrad nodding off strapped in his seat and Dick Gordon hanging out the door.

Finally they tried the tether exercise, backing away from their Agena and trying to set up a slow rotating motion that the theory said would eventually damp out. It didn't go as planned. The long tether whipped and sawed in slow motion. At one point it looped on itself. Conrad couldn't find a place where the tether was taut and they could simply float along. Finally it reached some small stability. Conrad and Gordon hoped to feel a small force of gravity as they rotated; they didn't. After three hours, they jettisoned the tether.

Gemini XI was a working mission. They dropped away from their Agena, got a short night's sleep, did still another rendezvous exercise, and finally finished up. Only one Gemini mission was left.

That one went to Jim Lovell and Edwin E. "Buzz" Aldrin. It was another success, crammed with learning experiences. They did a two-orbit rendezvous, filling out the chart so that Gemini had now done one-, two-, three-, and four-orbit rendezvous, each requiring different techniques to get there. Their Agena had a rocket engine problem, so they didn't get the high-altitude rides experienced on X and XI. But Aldrin's space walk finally gave an astronaut experience using tools in weightlessness, and he hooked up another tether to repeat the XI test. This time, after Lovell practiced the setup maneuvers with the benefit of Pete Conrad's tutelage, it worked almost as planned, and was logged in as a station-keeping method that might be used on far-future missions to save fuel.

Lovell brought Gemini XII to perfect splashdown, less than three miles from the aircraft carrier *Wasp*. Gemini was over. It was November 15, 1966.

Apollo was only months away. The first crew was long into training and getting ready to fly in early 1967.

Gus Grissom, Roger Chaffee, and Ed White watched the end of Gemini with keen satisfaction. Then they turned back to the job of meeting their own date with fate.

In the twenty-one months of Gemini, sixteen American astronauts went into space. Four of them—Tom Stafford, John Young, Pete Conrad, and Jim Lovell, all members of the Second Nine—flew twice. Of the sixteen, only Gordon Cooper would not be assigned to an Apollo mission. He was tired and it was time to move on.

They'd done rendezvous. They'd done docking. They'd fired Agena engines to go into high orbits. They'd done multiple space walks and worked out techniques to combat the fatigue and stress of working in a pressurized space suit while weightless. They'd flown long, up to fourteen days, and came home healthy. Every manned spaceflight record belonged to the United States, except orbiting a woman and putting three men in a can and shooting them around the world. America had not been first on some of the records. It had just been best.

In Houston, Bob Gilruth built an efficient Manned Spacecraft Center operation that ran Gemini while simultaneously getting ready for Apollo moon flights. He'd given great authority to Chris Kraft, who'd wrung out the bugs in a new control center and put teams of flight controllers through the sweat of training and the rigors of running missions twenty-four hours a day for up to two weeks; to Deke Slayton, who took iron-man control over picking astronaut crews and training them; and to Max Faget, who was overseeing development of the two Apollo spacecraft.

In the same twenty-one months, the Russians did nothing in public.

The workload on everyone both inside NASA and at its aerospace contractors—executives, engineers, astronauts, families, even the reporters assigned full-time to the space beat—grew beyond their comprehension and would accelerate to new levels of adrenaline-stoked intensity and exhaustion in Apollo.

It happened to everyone. The astronaut families had it just a little bit worse.

"Nobody thought about what life would be like afterwards," Jan Armstrong said with the benefit of thirty years' hindsight. "Nobody had time to even think about that. The guys would come off one flight and go right onto another."

George Mueller, the NASA associate administrator for manned space-flight at the time, was no more popular with some of the astronaut wives than he was with some members of the press. He'd have semiannual teas for the wives and give them a pep talk. His message was always the same: "Now, girls, you keep up the good work. Keep a stiff upper lip." Jan Armstrong married Neil in college, after his Navy service. She was not a military wife, the only one who wasn't. She thought Mueller's way of talking down to the wives was insulting.

Rene Carpenter was a military wife, but her independence streak ran over the edges of military protocol too. During Mercury, she'd do a routine that lampooned the neat little boxes into which astronaut wives and kids were supposed to fit. "Every woman has her own identity," she'd say, "and she's probably had to take a tranquilizer pill to step out in front and the inter-viewer says, 'Oh, yes, I want to introduce you to the lovely photogenic family of Commander Westlake and Mrs. Westlake were you happy, thrilled, and proud?'

" 'Well, I guess I was . . .' "

There was bitter truth in Rene's sarcastic humor. The wives played their role. They were strong women, beautiful women, the kind men dream about who can handle the situation, *any situation,* without breaking down or both-ering hubby with the details.

"We didn't burden them with the trivial things," Jan Armstrong remem-bered. "We had the pressure on us to make do. All the gals were busy raising their kids. The guys blew in and out." They lived with the knowledge that some of the husbands, or many of the husbands, were presented with too-frequent sexual opportunities and that they too frequently gave in to philan-dering urges. They lived, too, with the knowledge that their husbands could die tomorrow. It was part of the job.

They learned to live with sudden demands on their own lives too. Their neighbors were moon people too, working for NASA or for a contractor. "Something would come up and the gal across the street would take the kids," Jan said, "or I'd take hers. Nobody ever said no, nobody ever ques-tioned it."

The *Life* magazine contract gave the astronaut families some protection from the rest of the press. "We had no idea what was ahead," Jan said. "The *Life* people were wonderful to work with, they didn't try to take advantage. But they taught us how to deal with the press."

When they had to, usually after a mission, the appropriate wives would appear at a press conference where the full field of reporters could ask questions. "We'd joke about it," Jan said. "We were always 'proud, pleased, and happy.' That was what the world wanted to hear. They didn't want to hear that it was a helluva week at home. Junior has chicken pox and somebody broke his arm."

Jan Armstrong remembered a typical weekend as an astronaut wife. "The guys would come home late Friday, they'd flown in on 100 percent oxygen, we'd sit down for a late dinner, maybe a glass of wine, and then he'd be asleep." On Saturday morning, the astronauts would be in the office catching up on paperwork and talking with the other crews. "On Sunday, they'd be gone again," Jan said. "Maybe there was time to do something Saturday night. Maybe."

When their husbands came back from space, lives took on a surreal quality. There'd be parades and politicians and maybe dinner at the White House and a foreign tour. Women who'd been just wives and neighbors suddenly needed the right clothes, a week or two of baby-sitting, some knowledge of protocol. They were on their own to find clothes money in the budget and to find a neighbor or another astronaut wife to take the kids. Somehow it always worked out.

The "guys" promoted that kind of thinking. "We were given a lot of assurance by the guys that it was just a job," Jan said. "Maybe we were brainwashed. Deep down, we knew it was dangerous. But we had to be stoic, and we were in private too. Most of us came from the test pilot world. We'd lost a lot of friends. When Kinch [Iven Kinchloe] went down, it was devastating. When Joe Walker was killed, it wiped me out. Gracie was my best friend."

In the test pilot world, and in the Gemini and Apollo world too, good men were lost, and too often publicly.

Gemini was the bridge to Apollo. It was a magnificent accomplishment and remains a monument to American technology and dedication. There is a strong argument that the space race ended during Gemini. After Gemini, all that remained was for America to forge on and cross the finish line.

The astronaut wives had more exposure to the press and lived closer to death, but it wasn't that much different for all of the moon families, tens and hundreds of thousands of them. The husbands worked sixty, seventy, eighty hours a week; the wives ran the household and raised the kids. They all knew why and they believed.

A few men would get to the moon. A huge supporting cast, many of them women and children, helped them do it.

With Sergei Korolev dead, the Russian half of the space race fell into confusion and dissension. His successor was Vasily Mishin, but it took five months of bickering and infighting to make it official. Korolev's bureau had critics when he was alive, and at least one other bureau, headed by rival Vladimir Chelomey, was developing spacecraft and rockets for a lunar program.

Chelomey was more of a scientist and theoretician than Korolev, whose best talent was managing complex technical programs. But Chelomey knew how to get work. His bureau hired the best people it could find. One of them was a skilled guidance expert named Sergei Khrushchev, the premier's son. An important bureau assignment in the early 1960s was to develop a manned spacecraft and its rocket to do a mission around the moon and back. It wouldn't go into lunar orbit, but it would be a major technical and scientific feat.

At the same time, Korolev's assignment was to develop lunar landing hardware and the huge N-1 rocket the mission would need. It wasn't Korolev's only job. He was directing the Vostok and Voskhod programs, selecting cosmonauts for missions, working on lunar probes, on Mars and Venus probes, and on early spy satellites. Most of the moon and planetary probes failed; their rockets blew up, their guidance systems missed the targets, or the hardware simply went dead out there in space and was never heard from again. The Chief Designer and his bureau were spread thin.

That didn't stop Korolev from taking on the assignment in 1963 to develop a new rocket and a multimanned spacecraft called Soyuz. The Chief Designer and his chief rival both had Khrushchev's ear, but it was Korolev who masterminded the propaganda victories of Sputnik and the cosmonauts. Chelomey's circumlunar mission consumed time and energy, but never got to a serious stage. Even the lunar landing was not a serious program until August 1964, when Khrushchev pushed it through the Communist Party Central Committee and it was made a national objective. By then, America had a three-year head start, with the Saturn booster and the two Apollo spacecraft not just designed but being built and in early testing. Korolev had only drawings and designs. Even with party approval, a cosmonaut lunar landing did not get highest priority. Arguments raged between competing factions over the direction of the Soviet program, with arguments for a space station gaining currency. Korolev may have known that it would take a miracle to beat the Americans to the moon.

With Khrushchev deposed in late 1964, the infighting between space factions only got worse. It took a year, but Korolev had Leonid Brezhnev's attention and Chelomey did not. All lunar programs went to Korolev's bureau by the end of 1965.

While Gemini ascended, his engineers concentrated on the Soyuz program, with its multiman crews and docking capabilities in Earth orbit. One Soyuz docked with another could, in fact, be called a mini space station. Other parts of the Korolev bureau, with lesser emphasis and budget, worked on the N-1 rocket and lunar spacecraft.

That was what Vasily Mishin inherited in May 1966, after five months of confusion. The practical side of it should have been clear. It was one thing to throw a man in a can into orbit and reap headlines. It was quite another to send that man to the surface of the moon and bring him home again. There is an old joke that applied to Mishin's predicament. He had two chances to beat the Americans to the moon: slim and none. And Slim just left the room.

12

"Fire . . . I smell fire." The voice was unidentified.

"Fire in the cockpit!" Ed White's voice was urgent.

Gus Grissom was already grabbing for the ratchet tool that had to be plugged into a slot in the hatch and then turned, and turned, and turned many more times to get the hatch open.

"We're on fire . . . Get us out of here!" Roger Chaffee's voice was shrill, his agony apparent.

"There's a bad fire in the spacecraft!" The voice was choking, almost unintelligible.

There was silence on the voice loop. It had been just seventeen seconds from the first transmission to the last. Horrified launch controllers, Apollo managers, and astronauts including Deke Slayton watched on television monitors as fire flared inside the Apollo 204 command module, grew quickly bright as a sun, and exploded through the side of the ship.

At the 218-foot level of Launch Pad 34 at Kennedy Space Center, the explosion as the pressure hull ruptured threw rescuers against the wall. Concussed and ignoring their own burns, they tried to pull off the spacecraft cover, get to the hatch, rip it open in the dense, choking smoke. It took five and a half minutes. When the smoke cleared enough to see, Roger Chaffee lay dead on his couch. Gus Grissom and Ed White, dead in space suits that were charred and melted together by flames, were jumbled under the hatch.

Gemini had ended two months earlier. On January 27, 1967, Apollo stopped.

If the Apollo 204 tragedy happened today, the Apollo program might have been canceled. The space race could have ended, or gone eventually to the Russians. When the space shuttle Challenger exploded in 1986, there was a prolonged, acrimonious investigation, but the program got back on track. Something similar at the end of the twentieth century, with a more fragile citizenry and even more fragile political leaders, could stop manned spaceflight altogether.

It didn't happen in 1967. Times were different. The people and its government were stronger. Only a month before he died in a fire that might have been avoided, the normally underspoken Gus Grissom forewarned the world in an interview with Howard Benedict of the Associated Press: "If we die, we want people to accept it. We are in a risky business and we hope that if anything happens to us, it will not delay the program. The conquest of space is worth the risk of life."

The shock was complete. In Houston, reporters were called at home around 6 p.m. and told to come to the news center. The same thing was happening at the Cape. As the first reporter, a correspondent for *Time* and *Life,* arrived at the Houston newsroom, Jack Riley of public affairs was just putting down the phone. His eyes were tearing and going red. He could hardly talk.

"The crew," he said. "We lost the whole crew."

Joe Shea, manager of Apollo Spacecraft Program Office at the Manned Spacecraft Center, was on a National Airlines flight inbound from Florida. His seatmate was Ben Cate of *Time,* and the magazine was ready to put Shea on its cover. For a while that afternoon, Cate thought he would be flying home alone.

Both Shea and Deke Slayton had hoped to be inside the command module that day, watching the procedures and looking for problems. There was just enough room in the well beneath the astronauts' feet for two men to crouch or squat. Gus Grissom vetoed the idea; there wasn't the right kind of communications gear for Shea and Slayton to be truly involved and they'd just be a nuisance anyway.

Grissom, Chaffee, and White were picked by Deke Slayton to fly the first manned Apollo mission. It was scheduled for February 21, an open-ended flight that could run as long as sixteen days while they shook down the command module and the large cylindrical service module that carried consumables like oxygen, water, and rocket fuel, and which housed the big rocket engine that was designed to get astronauts home from the moon.

If Apollo 204—201, 202, 203 had been unmanned flights to test the Saturn 1B rocket and the command and service modules—was a good mission, it looked strongly like Americans could walk on the moon in 1968. The fire changed all that, and brought new insight into what they were dealing with to NASA leaders.

"There was enough wrong with the spacecraft that without the fire, we might not have made Kennedy's deadline at all," Chris Kraft said much later. "We'd have flown, found problems, taken months to fix them, flown again, found more problems, taken more months . . . We might not have landed on the moon until 1970 or '71." Instead they tore into every part of the command module, reexamined the service module, went deep into the systems of the lunar excursion module—the "excursion" was dropped when stiff-necks at NASA headquarters thought it was too frivolous to be part of the name—and made thousands of fixes major and minor.

President Lyndon Johnson ordered NASA to conduct the investigation internally and to do what it decided was right. There was no outside board of experts, no second-guessing by people whose main qualification was to possess twenty-twenty hindsight. The insiders did enough of that themselves. Congress stood back, with the House and Senate holding off on public hearings until NASA had something concrete to report.

Bob Gilruth sent astronaut Frank Borman to the Cape to help with the investigation. The night after the fire, a group of them were in the Vanguard Lounge at Cocoa Beach. It had been an evil day and they were maudlin about losing their friends, angry that they'd somehow allowed it to happen. So they were tossing them back, and Borman, a Boy Scout cast in John Glenn's image, got drunk. Paul Haney and Joe Shea watched him slowly close his eyes and lay his forehead into a bowl of shrimp peels. "That's not like Frank at all," Haney said, but no one argued with his right to pass out.

They never found out whom to blame. It was no one and everyone. The cause of the fire, some spark or short circuit or other ignition source, remained a mystery. They did quickly realize why the fire burned so rapidly and intensely. There were two reasons.

The test that day was called "plugs out" and involved sealing the command module tight, then running it on its own internal electrical power system. The atmosphere in the cockpit was 100 percent oxygen. The Russians used an oxygen-nitrogen mix, but that had its own drawbacks for space walks and two tanks of breathable gases weighed more than one tank. All American spacecraft ran on 100 percent oxygen. But with the cockpit sealed, it was pressurized above normal sea level pressure, to 16.7 pounds per square inch.

Fire burns fast and hot in pure oxygen at normal pressures. At a higher pressure, it burns even faster and hotter.

The second reason was that the cockpit was loaded with flammable objects. Flight plans and other necessary documents were printed on regular paper. The space suits were made of flammable nylon. So were the couch covers. The cushions in the couches were polyester foam. The cooling system was filled with ethylene glycol, which burns rapidly in pure oxygen and with dense, acrid, toxic smoke. There was Velcro attached to almost every free surface. Astronauts love Velcro; in zero gravity, it gives them a way to quickly attach something—a camera, a food pouch, a harmonica—to a handy place where it won't float away.

Within a month, the investigating team had a long list of recommendations, and work started immediately to implement them on future Apollo spacecraft. Among the first, they insisted that flammable materials be replaced wherever possible and that an easy-open hatch be designed. Both were done. Assembly and fabrication procedures were tightened at the North American Rockwell plant that was building command and service modules in Downey, California, and at the Grumman Aircraft plant on Long Island where lunar modules were under construction. Through the investigation, the NASA review board had found hundreds of instances of sloppiness, and these and others found by the contractors were remedied.

The pure oxygen atmosphere was criticized by everyone. In the future, ground tests and the launch itself would see an atmosphere of 60 percent oxygen and 40 percent nitrogen, in which fire was much less likely. On the way to space, the atmosphere in the cockpit would be replaced with pure oxygen. Tests showed that fire in zero gravity was far less likely and more easily contained than fire on the ground.

There were hearings in both the House and the Senate, with much criticism leveled at NASA for allowing the fire to happen. Some congressmen, particularly William F. Ryan, a Democrat from New York, were outspoken in their condemnations. Frank Borman defended NASA in personal testimony. "We are trying to tell you that we are confident in our management, and in our engineering, and in ourselves," he told a House committee. "I think the question is really: Are you confident in us?" Eventually Congress acceded.

At the Manned Spacecraft Center, moods ranged from grumpy on a good day to bleak and depressed most of the time. In a board meeting of the Escape Velocity Press Club, directors decided to do something positive. The sixth anniversary of Al Shepard's historic suborbital flight was coming. Nobody celebrates sixth anniversaries. The club decided to throw a banquet honoring Shepard and to donate the proceeds to the just-announced Ed White Memorial Fund.

On the night of May 6, 1967, the banquet hall at a local hotel was filled with more than five hundred of the nation's senior space people. Aerospace industry executives flew in from California, New York, Florida, a half dozen other states. The front tables of honor were filled with Bob Gilruth, Chris Kraft, Wernher von Braun, Deke Slayton, NASA people from Washington, and a few reporters from the press club board. Editors from *Life* magazine sat with photographer Ralph Morse at a table farther back. *Time* correspondents were there, along with senior writers from the Associated Press, United Press International, Agence France-Presse, Reuters, and more. Television was represented with reporters from New York and Washington. It was a crowd, and the goal for the evening was to get them to lighten up.

The audiovisual contractor at the space center put together a movie, with musical help from record producer Mickey Kapp. It was a pseudo-biography of Al Shepard, who was making a reputation as an astute investor as well as chief of the astronaut office. Wally Schirra was the narrator, and the film featured a series of airplane crashes from the pioneer days as well as footage of numerous rockets exploding just off the launch pad.

The movie's title was *How to Succeed in Business Without Really Flying—Much.*

Shepard cracked up, and the audience was quickly laughing harder than any of them had laughed since January 27. Astronauts got into the act with a singing vaudeville routine that included Shepard. In the original Broadway show *How to Succeed in Business Without Really Trying,* singer Robert Morse stares at himself in the mirror and sings "I Believe in You." Mickey Kapp rewrote the lyrics, and Shepard stood there in front of his peers and sang "I Believe in Me." He wasn't bad.

When it was over, Wernher von Braun said a few words in tribute to Grissom, Chaffee, and White, then spoke of the inherent and accepted dangers of test flying and spaceflight. "This reminds me of what Bill Allen, the president of Boeing, said the other night," von Braun said with a catch in his voice, ". . . we're not in the business of making shoes." There were few dry eyes in the room and a lot of nods.

Then it was Shepard's turn to speak. He was not the tongue-tied, embarrassed fellow later portrayed in an HBO television series, nor was he afraid to speak his piece. He'd thought carefully about his remarks for that night and he was eloquent.

"If the third unmanned Redstone flight had not been made, it is entirely possible that my flight could have occurred prior to Gagarin's and the temporary flush of victory would have been ours.

It was discouraging to all of us. Some of us in this room who helped make that decision were heartsick."

Until this moment, he had not looked directly at Wernher von Braun for the entire evening. Now he did and his words seemed conciliatory.

"We were so close to the event that the country needed.

"As I pointed out, a different decision could have given us the first flight—*and could just as easily have resulted in failure.* I supported that decision then and I still support it today.

"History now shows us that this month gave only temporary advantage to another political scheme. It exonerates the careful judgment of you men who placed success over temporary propaganda advantage.

"Our total space effort today is second to none. I want every chance for this country to be first in everything it does. And yet, if we should lose the race to the moon, say for example, by a month, we cannot be more than temporarily dismayed. The important thing is that our decision is to pay for and project the technology that this country must have to maintain this leadership and to prove, and to improve, our domestic standards. My point is that we will be remembered in fact for *how* we did it, and not *when* we did it. And that our decisions next week and next year were made with the same courage and convictions displayed in 1961."

By now, Al Shepard had been interrupted with applause several times. It was prolonged after that statement. He held up hands for silence. "I would like to close with a few words about the recent accident." The noise level in the room fell to an absolute hush. Not a coffee spoon or a cleared throat was heard.

"Much has been said about the loss which we all felt so keenly, and appropriate memorials such as the one we support tonight have been established. May these types of efforts continue, especially to help us in our future rationale. Much has also been said about the cause and effects of the fire. In this case, perhaps too much. I am not from New York State. I have no current desire to perpetuate a political career. And I say without malice that the report of the investigation committee by itself is sufficient retrospect.

"All of us here tonight jointly share the responsibilities for the human frailties which are now so apparent and for the insidious combination of materials and equipment which was so devastating in their behavior. We jointly share the responsibility for future prevention of similar circumstances. It is not inconceivable to me that even a non-technical reporter could spotlight some of the deviations in planning and management."

He looked now at a *Time-Life* correspondent seated just below the rostrum. Then he got to his point.

"The time for recrimination is over. We have digested enough historical evidence. There is much to be done. Morale is high. Vision is still clear.

"And I say, let's get on with the job."

The standing ovation was long and thunderous. Everyone knew: It *was* time to get on with the job. And they did.

Joe Shea was replaced as Apollo manager and moved to a meaningless job at NASA headquarters. He soon resigned, blaming himself for the fire and wondering if he could have saved the crew if he'd been squatting there inside the cockpit when it started. George Low replaced him in Houston. At North American Rockwell, a few executives in the Apollo chain of command were moved to other jobs. Jim Webb stuck around until just before the 1968 election, then resigned as NASA administrator.

Bob Gilruth pushed the Apollo spacecraft engineering team to new limits, and in just seventeen months, they were ready to fly again.

Twenty years after the fire, Lee Atwood, president of North American Rockwell from 1948 to 1970, wrote a long and thoughtful recollection of the fire and its causes. His postscript to that recollection put an important perspective on what it means to be involved in test flying and spaceflight.

It is, of course, not possible to know, but if the question had been put properly to any of the top Apollo executive (including the writer) —e.g., "Did you know that the astronauts are being locked in with all that electrical machinery and the spacecraft is being inflated to 16.7 pounds per square inch with pure oxygen?"—I believe a whistle would have been blown.

But what about the Congressional investigators' favorite: "You knew or should have known"? "Should" is an interesting word. It lays on an obligation of diligence, perception, action, and rectitude with little reciprocal responsibility. It follows most of us all our lives, but fortunately, it seldom stops the clock.

After all, organizations are large, and we all take a lot for granted. But "should" is pesky. It seems to skulk around just outside the light of the campfire, and sometimes when the moon is full, its shadowy profile can be seen, along with the misty images of long lost airplanes.

By heaven, methinks it were an easy leap,
To pluck bright honour from the pale-faced moon . . .

Between Frank Borman's plea to Congress and Al Shepard's eloquence at the press club banquet, Vladimir Komarov died on the first Russian space mission in more than two years. It was a bad beginning for Vasily Mishin's stewardship of the Russian lane in the space race.

Soyuz 1 quickly followed two successful unmanned test flights in March and April 1967. The Soyuz could carry two or three, or maybe more, cosmonauts in some kind of comfort. It was designed with its own rocket engine for rendezvous maneuvers in space, and it could dock with another Soyuz or with an unmanned cargo carrier. It was Komarov's second flight. He was the first cosmonaut to go back into orbit. Near the end of his first day, the Soyuz control system began acting up. Mishin decided to bring him down after his nineteenth orbit.

The problem came after a perfectly aligned and executed retrofire maneuver. Soyuz 1 plunged on course through the atmosphere and emerged from the fireball of reentry intact. Komarov was on the radio reporting normal spacecraft functions when the Soyuz parachutes popped out and their lines tangled. The parachutes didn't deploy. Soyuz 1 crashed to the ground, killing Vladimir Komarov instantly. NASA detailed Frank Borman to represent it at Komarov's funeral. When his plane stopped to refuel in Copenhagen, the Russians informed him that he would not be allowed to attend. Borman returned home. The Russian manned space program came to a complete stop.

Now both runners in the space race were in trouble. The Russian troubles were worse.

There was a big Saturn rocket and a little Saturn rocket. Wernher von Braun had done his job producing them. Trying them out in a series of unmanned test flights would tell how well he'd done it. The Saturn 1B, for a brief time billed as "America's newest and biggest rocket," was the little one. It was a two-stage booster standing 224 feet tall whose only purpose was to be the launch vehicle for unmanned Apollo tests and one or two manned missions. Its first stage was powered by eight rocket engines, burning a special kerosene fuel combined with liquid oxygen and each generating 200,000 pounds of thrust.

Called the H-1, the engines were testimony to von Braun's long-term genius in rocketry. They were bigger and more powerful, but their design was essentially the same as the Jupiter rocket von Braun had been firing since the late 1950s. The new technology was in their huge size, not in their inner works. Yet compared to the engines developed for the big Saturn V, they were dwarfs. The second stage, called the S-IVB, would be used on every Apollo mission. It had a single engine also generating 200,000 pounds of thrust. But its fuel was liquid hydrogen, and except on test stands, it would only be fired in space. It was a new engine. NASA planned three Saturn-Apollo tests using the 1B before trusting men on it, all in 1966 while the Gemini program was flying. The first came February 26, a few weeks before Gemini VIII. It was a simple mission: test the first and second stages by sending an Apollo command and service module combination on a long, high arc that would also test the spacecraft.

It turned into a paradox—a comedy of confusion yielding a successful mission. After a one-day slip for weather, the countdown was smoothly down to zero. At the press site a mile away, reporters with binoculars in one hand and a phone in the other prepared to tell editors that the big rocket was lifting off. It didn't. The rocket engines should have ignited at T minus three seconds, with liftoff at zero. Nothing happened and the big countdown clock stopped exactly at zero.

Confused reporters at the press site didn't know that controllers in the firing room at Kennedy Space Center and in a new control center designed just for Apollo 950 miles away in Houston were just as confused. In the history of space launches, there had never been a "hold" at T minus zero. It couldn't happen.

Within minutes, they fixed the problem—a faulty nitrogen bottle on the pad—and tried again. This time the countdown got to T minus five minutes before it stopped again. Something was loose on the spacecraft's protective cover. The word came from Houston. The mission was scrubbed. Most reporters milled around, waiting for an explanation. A few hopped into cars and headed for the airport thirty miles away in Melbourne. One of them was Jules Bergman, science editor for ABC Television. His report on the launch was logged in for the ABC evening news show that night. He glanced at his watch, saw that he could just make an 11 a.m. flight to New York, and roared away in his rented Chrysler convertible.

Suddenly the press site loudspeakers squawked. "Attention, please. Attention, please. Upon review, Houston has unscrubbed the scrub. Repeat, Houston has unscrubbed the scrub. The countdown will resume shortly at T minus five minutes." In the intervening minutes, a launch technician had gone up the elevator to the top of the Saturn rocket, slapped some tape on a loose piece of the spacecraft cover, and come back down. Houston spacecraft experts decided that the tape was good enough to hold through launch, and for the only time in space history, a scrub was unscrubbed.

Jules Bergman was twenty miles away with the top down on his convertible when he heard the roar of rocket engines behind him. It only took a glance over his shoulder for his heart to stop. The first Saturn 1B, America's biggest rocket, was rising majestically on an enormous pillar of flame and an ear-pounding crackling rumble. Slowly it tilted oceanward, accelerated under 1.6 million pounds of rocket power, and climbed toward space. It was 33 miles up when a bright flash visible to the unaided eye marked the shutdown of the first stage and the activation of the S-IVB stage. By the time Bergman got back to the press site, the mission was all but over.

The S-IVB pushed Apollo to 264 miles high. It coasted up another 39 miles before arcing back toward the Atlantic. At that point, the service module's big engine fired for three minutes to ram the spacecraft hard into the atmosphere. After a quick wait, it fired again for another ten seconds; that proved that it could restart in space. Only thirty-seven minutes after it left the launch pad, the Apollo command module came down perfectly under parachutes and was pulled aboard ship. The mission was declared a success. The few problems seen weren't destructive and would be fixed.

Jules Bergman and his cameraman climbed back to the roof of their press site studio where a desk and chair waited behind a white railing. For the next ten minutes, Bergman chattered excitedly into his microphone, staring skyward with binoculars and pointing while the camera captured his act. That night on ABC, it looked almost real. The editing room had adroitly interposed shots of the Saturn rising with Bergman pointing and describing its "magnificent climb" toward space. Reporters in the Mousetrap Bar at Cocoa Beach that night watched his performance over beers and boilermakers, and got a great laugh.

The next Saturn flight came in July, without an Apollo on board. Scientists still knew little about how fluids behaved in zero gravity. The mission put the hydrogen-fueled S-IVB stage into orbit, then let it orbit the Earth while they watched its liquid hydrogen tank by television. There was some sloshing, but it wasn't serious. NASA was ready to say that the S-IVB could indeed shoot Apollo out of orbit and send it to the moon. A final unmanned test came only seven weeks later. This time the Saturn and the S-IVB put an Apollo on a long trajectory to the Pacific. Again the service module engine fired, now four times, and the command module was recovered in the water. The Saturn 1B and the Apollo command and service modules were officially rated and ready to take men into orbit around the Earth.

The fire at Pad 34 five months later changed everything.

It took only six months after cosmonaut Vladimir Komarov's death for the new chief designer—it was hard for even his own people to think of him that way after the Korolev years—to get some legitimate attention, and briefly tables were turned.

In the United States, Apollo was being torn apart and rebuilt. It was still a year before astronauts would take it into orbit, and nobody was saying when a lunar mission might be attempted. The quiet confidence of earlier years was missing.

The Russians tried to step into the gap. Over three days in late October 1967, Vasily Mishin put two Kosmos ships into orbit, and this time, without benefit of a human on board, the Russians did a rendezvous and docking. It was a major accomplishment that attracted world attention again and would become a standard Russian practice in the coming years of space stations. But it did nothing to advance Mishin's standing in the space race to the moon. For that he needed men in space. Like Apollo, that was again somewhere in the future.

But more unmanned flights of Saturn were not. The first all-up launch of a Saturn 5 rocket was a literal wake-up call when it shook its world at 7 a.m. on November 9, 1967. After the fire, they'd renumbered the missions. The fire became Apollo 1. There was confusion about numbering the three Saturn 1B flights. No matter. Somebody made a ruling. It was arbitrary and they never explained it. The first Saturn 5 mission was Apollo 4.

Saturn 5 is huge. Three of them remain, one on display at the renamed Johnson Space Center in Houston, another at the Marshall Space Flight Center in Huntsville, Alabama, and a third at Kennedy Space Center, Florida. They must be seen to be believed.

This was the rocket that would take astronauts to the moon. It was 364 feet tall, including the Apollo spacecraft at its top. Its first stage, the S-1C, was 138 feet tall and 33 feet in diameter. A cluster of five F-1 rocket engines protruded from its bottom. Each engine produced 1.5 million pounds of thrust—7.5 million pounds altogether. It was the largest rocket ever built, and it still could trace its origins back to Wernher von Braun's early designs.

The kerosene and liquid oxygen alone in the first stage filled fifty-four railroad tank cars. Each of the five fuel pumps had about the same horsepower as six diesel train engines. The explosive power of Saturn 5 was so great that NASA moved the press site back from one mile away to three and half miles from the launch pad. The launch control center nearby was fitted with steel shutters which would shut automatically if the rocket exploded. That fact did not escape notice of reporters, who had nothing between them and the launch pad but three and a half miles of air.

The second stage, the S-2, was also 33 feet in diameter, but only 81.5 feet tall. That still made it taller than the Redstone rocket that put Al Shepard and Gus Grissom onto suborbital spaceflights. It also had five engines, but smaller. These used liquid hydrogen and oxygen for fuel and produced about 230,000 pounds of thrust each. So even the Saturn 5's second stage had more than 1 million pounds of thrust—more than three times as much as an Atlas rocket.

It only took thirty-four railroad cars to carry the hydrogen and oxygen for the second stage.

The third stage was the faithful S-IVB, already flown three times on test flights. Here the overall Saturn stack tapered down to a 21.7-foot diameter. An electronics and computer ring called the "instrument unit" sat atop that stage. It was the automated in-flight control center for Saturn.

Above the third stage, the stack tapered again. A lunar module was stored inside there. On the first Saturn 5 flight, it was a dummy, just a lump of equal mass and distribution. The service module, and then the astronaut-carrying command module, sat atop the lunar module's "garage." There were no astronauts on board on that November morning. But it was in many ways a practice moon mission, and when the day was over, people were still shaking with excitement.

At the press site, packed with more than a thousand reporters and television people, the new loudspeakers were crisp and clear as Jack King of the Kennedy public affairs office counted down to the moment of launch. Nobody really knew what to expect.

"T minus twelve, eleven, ten, nine, ignition sequence start, five, four . . ."

Hearts were thumping at the press site loud enough to be heard over King's voice.

"We have ignition!"

A giant cloud of smoke and fire erupted from the base of the rocket so far away, yet so huge that the distance was deceiving. The only sound was King's voice.

"All engines ARE RUNNING! WE HAVE LIFTOFF! WE HAVE LIFTOFF AT 7 A.M., EASTERN STANDARD TIME!!!" He'd started calm, but the moment overpowered him.

It took nine seconds to start moving, nine seconds during which the F-1 engines burned 135 tons of propellant and built up to full thrust. Almost so slowly that it surely had to be wrong, the Saturn 5 climbed against the launch pad and its towers. Its enormous flame lengthened as it climbed, always touching the base of the pad but extending up to the climbing giant.

There was still no sound from the monster. At sea level, it would take sixteen seconds for the noise to cross to the press site. Pressure waves moving through soil and rock are much faster. The ground beneath the reporters' feet began to tremble. It shook. It amplified. Reporters felt the Saturn 5 long before they heard it. All one thousand of them were screaming and yelling. "Go! Go! Omigod! Omigod!" No one was ready for this.

A few yards away, ceiling tiles in Walter Cronkite's too-flimsy studio began to fall around him, live on CBS Television. The walls of the NBC studio next door shook and rippled. Jules Bergman at his ABC rooftop vantage point could hardly speak as the floor beneath him shivered violently. Reporters in the bleacher seats, a sturdy metal structure with a heavy steel roof, scrambled to get away. The I-beams were shaking and the steel roof sheets rippled like tinfoil. There was still no sound except the screaming and the roof rattling and the loudspeakers.

"THE TOWER HAS BEEN CLEARED!" Jack King shouted. That was the cue for control to switch from the Cape to Houston. The next words from the loudspeaker came from the mission control center in Houston.

"Roger, Jack, tower clear fifteen seconds out." Nobody heard the last half of that sentence. They had to read it and everything else that came out of mission control for the next several minutes when the transcript was typed and laid out on tables. Nobody heard it because the sound of the Saturn 5 had just arrived.

It was unique and overpowering. At first, for a half second, it was the crackle of an immense fire. The crackle was overlaid with a thunder that didn't stop. The thunder was overlaid with the roar of a dozen hurricanes. The crackle came back in orchestral counterpoint, arguing with the hurricanes for dominance. Air pushed against faces and ears with a pressure that wasn't wind, just a steady shove. Reporters leaned into the shove and craned to watch that omigod rocket pitch over on cue to a straight east heading.

Like a symbolic marker pointing the way to the moon, the Saturn 5's fire tail spread out as it climbed into thinner air. It became a perfect red-yellow arrow racing across the sky, climbing and accelerating. The higher it went, the larger the arrow became. It had to be an omen and it was there every time a Saturn 5 was launched.

Across the state in St. Petersburg, well north to Jacksonville, south as far as Palm Beach, people saw the arrow in the sky and some heard a distant rumble that wasn't thunder. Then it was into the high clouds and gone. At the press site, reporters looked at each other and shivered. The shaking and the noise were gone. The memory would never go. *Omigod!*

It was a perfect mission. The command module had many of the modifications demanded by the fire, including a new hatch and an uprated heat shield. This was a test of structural integrity and of Apollo's ability to come home from the moon.

On its second orbit of Earth, the S-IVB stage fired again on a maneuver that would become familiarly known as TLI—translunar injection. This time it wasn't a full-power firing. It pushed the Apollo spacecraft to 10,696 miles, where it separated from the rocket stage. Then the service module's big 22,400-pound engine fired to raise the apogee to 11,234 miles. Over the next four and a half hours, it coasted and soaked in the solar heat and utter cold of space while instruments measured how it held up.

It held up fine. Then on command, thrusters fired to turn it around, nose pointed back toward Earth. The big spacecraft engine fired again to drive it down hard. Its speed built up to nearly 25,000 miles per hour, almost exactly how fast Apollo would be going when it returned from the moon. It became a fireball in the atmosphere; the temperature on its heat shield was five thousand degrees, nearly two thousand degrees hotter than either Mercury or Gemini endured.

The U.S.S. *Bennington* waited in the Pacific. Nine hours after liftoff, the Apollo command module's parachutes billowed and it splashed down just ten miles from the recovery ship.

Some NASA officials declared that America was ready to go to the moon. They were too optimistic. There was still work to do.

Next up was the lunar module (LM). Apollo 5 used a faithful Saturn 1B to put a legless LM into orbit and try it out. It didn't need legs because it wasn't going to land on the moon.

The lunar module was a strange-looking machine. Its angular lower section sprouted those legs for a moon mission, and a big rocket nozzle pointed straight down from its center. This was the descent stage and would be used to settle on the moon, with the throttleable rocket engine acting as a brake. It was the first rocket that an astronaut could control like a car engine, increasing or decreasing its power as needed.

The top part was angular too, but more rounded. This was the ascent stage and it carried the astronauts. There were no seats; they stood up, except when they were on the moon and curled up on the floor to sleep. Two triangular windows canted down to give them a good look at whatever was outside. A squarish hatch opened onto a flat area they called the "porch." There were four legs and one of them centered on the porch. A ladder ran down the leg, and at the leg's bottoms were big dishlike saucers that were the landing pads. The bottom part of each was built like an automobile shock absorber: land too hard and it stroked inward. Hidden under the ascent stage's middle was another rocket engine. This would be fired to launch off the moon and to rendezvous with the command and service modules. The descent stage would stay behind, serving in effect as a disposable launch platform.

This was what went into space on January 22, 1968—a legless LM. It got high marks for its performance. Once in orbit, the three panels of its garage opened and the LM popped out for a solo run around Earth. Three times ground controllers sent commands to turn on the descent engine, running up its throttle to maximum thrust on the second two burns.

Next they did an abort maneuver that separated the LM's two stages by firing the ascent engine. On the second firing, attitude thrusters overreacted when ground controllers left the autopilot turned on. They fixed the problem and logged the results as a stress test which showed high-quality system performance. A third firing went until the ascent stage ran out of fuel. The stage was brought down deliberately near Guam and disappeared in a fireball no one saw. Analysts pored over the flight data until a recommendation was obvious. A second unmanned LM flight could be canceled. In one unmanned mission, America had taken two steps toward the moon.

While they got ready for the next Saturn test, Radio Moscow made a sad announcement. Yuri Gagarin had been killed in a plane crash. He was rumored to be the head of their lunar landing team, if one even existed. His ashes were interred in the same Kremlin wall as Sergei Korolev's. If his death had an impact on Russian moon plans, it wasn't obvious. There was no impact on Apollo at all.

But Apollo 6 in April 1968 looked like a full step backward. The second unmanned flight of Saturn 5 left watchers as shaken as the first. But this time, TV studios at the press site and the bleacher structure had all been reinforced. The ground and the reporters still shook. The buildings didn't. Then dismay. As the five F-1 engines in the first stage fired, something wasn't right. There were fluctuations and the bouncy effect called pogo began to rattle the command module high above. The jarring and shaking was too powerful, more than doubling the design limits. It stopped when the first stage fell away, but it was a problem that had to be solved.

Then two of the five engines in the second stage inexplicably quit. The three remaining engines fired until the hydrogen/oxygen propellant was gone, but it wasn't enough. The third stage took over, but it couldn't put the unmanned Apollo into the right orbit. Instead of a circular orbit at 100 miles up, Apollo was in an elliptical 111-by-228-mile orbit. When controllers attempted to restart the third stage for a translunar injection maneuver, it was dead. Nothing happened.

They separated the Apollo craft from the silent stage and quickly worked out a new plan to use its big engine for another test of reentry at lunar speeds. It didn't get quite that fast, but was close enough. The spacecraft part of the mission was a success. The Saturn 5 part was not.

Before the Saturn troubles were solved, Neil Armstrong did it again. In a training exercise at Ellington Air Force Base only a few miles from the Manned Spacecraft Center, he came within two-fifths of a second from dying.

He was practicing the critical final moments of touching down on the moon, flying a specially designed lunar landing training vehicle (LLTV) in early May. It was his twenty-first flight in the weird flying machine. No other astronaut had his experience. Nicknamed the Flying Bedstead because of its resemblance to Granny's four-poster, the LLTV used a down-thrusting jet engine to remove five-sixths of the force of gravity. The astronaut sat at one corner of the bedstead to operate its controls. A typical training session meant a near-vertical climb to seven hundred or eight hundred feet under full power, then reducing the jet thrust to counterbalance gravity until it equaled what they'd find at the moon. For the remainder of the seven-minute flight, he used two lift rockets and sixteen small attitude-control thrusters to glide through a final descent, hover, and landing.

Armstrong had it cold for the first five minutes that day. Then suddenly the LLTV dropped vertically. Armstrong caught it, held it in a hover near the ground, then the craft shot up for two hundred feet. He fought it back to control, but then it began darting rapidly to the left. He stopped it again. In the next four seconds, the LLTV gyrated wildly over the Ellington runway. Its nose pitched sharply down, then steeply up, then it rolled right.

"Better get out of there, Neil," a ground controller shouted into the radio. The Flying Bedstead rolled sideways out of control as Armstrong activated its ejection-seat rocket. It was the first time a human had used that kind of seat. The Bedstead plunged to the concrete runway and exploded.

Armstrong's parachute opened just before his feet touched ground, then dragged him away from the fire. His only injuries were bruises.

The investigation board found that the LLTV's pressurized helium supply had run out early, caused by gusty winds that required Armstrong to use extra fuel, and by Armstrong himself forgetting to shut down the throttle to his lift rockets as he neared a landing. Under the conditions, it was a forgivable oversight. When they reviewed films of the accident and timed the events, they saw that if Armstrong had ejected two-fifths of a second later, his parachute would not have opened. Like Ted Freeman a few years earlier, he would have hit hard and died.

Space engineering can be as much detective work as technology when the parts that failed have incinerated in the atmosphere or sunk to the bottom of the ocean. So Wernher von Braun's Saturn engineers went into detective mode to find out why their second Saturn 5 had behaved so badly. There were three problems to solve: pogo as the first-stage engines were firing, the mysterious shutdown of two second-stage engines, and the third-stage failure to restart in orbit.

In Gemini, pogo had been traced to a partial vacuum developing in fuel lines; fuel flowed erratically, so the rocket engines stuttered and set up a bouncing vibration. They'd solved it by adding accumulators to pool fuel just before it hit the engine and standpipes to bleed off vacuum bubbles. Von Braun's rocket engineers put a Saturn 5 first stage on a test stand at Huntsville and isolated the problem to the liquid oxygen lines. The solution was simple: They added a system to inject pressurized helium behind valves just before the oxygen hit the firing chamber. When the engines fired, it worked just like a shock absorber and prevented the liquid oxygen from surging.

The second- and third-stage troubles involved the same kind of J-2 rocket engines. Nothing similar had ever happened during tests on the ground, so the engineering detectives deduced that it was somehow related to the conditions in space. But which condition? Again an engine was put on a test and fired. This time cameras and sensors monitored virtually every part of the rocket.

Almost immediately they spotted two problems. Liquid hydrogen and oxygen are extremely cold, but the interior plumbing in the engine heated up quickly. The detectives saw a layer of frost forming on the propellant

lines, protecting them from the heat. But in space there would be no frost. Temperatures were hot enough to burn through a thin-walled line. The solution was to strengthen the propellant lines, making them thick-walled enough to prevent burn-through. The second problem was a bellows inside the lines to help push the flowing propellants. In a vacuum test, it vibrated so violently that it broke. The solution was the same; with a strengthened propellant line, the bellows was simply eliminated.

Suddenly problems that threatened the entire Apollo flight schedule were solved, and in such a short time that NASA flight planners took a new look at the future. If Apollo 7 is a good manned flight in October 1968, why not send Apollo 8 to the moon?

Bob Gilruth's philosophy—it became the philosophy of the entire manned spaceflight team—was that every mission should move a program forward, building on previous missions and extending the capabilities of both the hardware and the men. In late 1967, Gilruth signed off on a plan developed in Houston. They argued it with NASA headquarters and it was approved. Under the Houston plan, if no major problems cropped up that required a repeat performance, it would take only eight Apollo missions to put men on the moon. Each was designated by a letter, A through G. Now three of the missions—the two A missions to test Saturn 5s and the command and service modules, and the B mission to try out a lunar module— were history. The rocket problems were solved. The biggest remaining problem on Earth was that construction of LMs for manned flights was running behind schedule by about sixty days.

The C mission, Apollo 7, was rapidly approaching. Wally Schirra was the commander, with rookies Walt Cunningham and Donn Eisele. They'd been the backup to the Grissom crew; now they'd be first in space with Apollo, using the smaller Saturn 1B to get there. Their job was to stay in orbit for up to eleven days and give the command and service modules a hard workout.

The D mission belonged to Jim McDivitt, Dave Scott, and Rusty Schweickart. They'd be first to ride a Saturn 5 and to take a lunar module along for a manned trial in low Earth orbit. But their LM had problems, and after it arrived at the Cape, it was being torn down and fixed. It wouldn't be ready to fly on the planned schedule of December or January.

Another crew was in training for the E mission. Frank Borman, Bill Anders, and Mike Collins would have the job of going to a high Earth orbit with the full set of Apollo ships in February or March 1969. There were no

crews assigned yet for the F and G missions, taking a LM into lunar orbit on F and then landing on G.

When he got the news that the Saturn 5 problems were fixed and that it was safe to put men up there on its next launch, Apollo manager George Low in Houston was already thinking. The Schirra mission was locked in and looked good for October. The McDivitt mission was a different story. They had a good Saturn 5 waiting, and their command and service modules were showing up almost clean in checkout. Did it make sense to have the men and the hardware sitting around for months with nothing to do except wait for a lunar module?

No, Low decided when the Saturn news came in, *it didn't.* He picked up the phone in early August and asked Chris Kraft to drop by Low's office in Building 2 for a quiet talk. When Kraft arrived, Low put it to him: Could Kraft's people put together a flight plan that would send Frank Borman's Apollo 8 crew to the moon before the end of the year?

Kraft lit up. He didn't just like the idea. He loved it. He went back to his office and began calling in the troops to make it happen. The next day Low went to the Cape to test the idea with the rocket people. Could they have the next Saturn 5 ready in time? He already knew the answer, but now it was official. Yes. They called it C-Prime.

Deke Slayton immediately shuffled his crews. The Borman crew, training for a high-altitude E mission, was handed the Christmas trip to the moon. Again if everything worked right, the E mission would be dropped from the plan. Almost immediately Slayton had to make another change. Mike Collins needed surgery to remove a bone spur on his spine; he wouldn't be ready for Apollo 8 and C-Prime. Slayton moved Jim Lovell from the backup crew into his slot.

Collins was a most unhappy astronaut at missing his chance to go to the moon, but the bone spur was painful. If he healed in good shape, Slayton promised, he'd be in line for another mission somewhere down the line.

Some of the Apollo 7 crew stories became classics. Wally Schirra was a sailor and loved his boat, a Cal-21 he named *Countdown.* He even made certain that his sail number was correct: 321. Al Shepard and Bill Dana decided to loosen things up and pulled a *gotcha* on the king of gotchas.

They found a mast identical to *Countdown*'s, then with a little help from

their friends, moved Wally's boat from its slip on Clear Lake into a nearby boat shed. Finally they planted the new mast deep into the mucky bottom of the slip; only a few feet showed above the water.

A call from the marina warned Schirra that something had happened to his boat. He roared the three miles from his house to the dock, saw only the tip of a mast where his beloved *Countdown* should be, and almost went into orbit without a rocket. Shepard and Dana were hiding nearby and couldn't control their laughter. *Gotcha, Wally!*

Apollo 7 came off the pad late. Sam Phillips had told a press conference two months earlier that launch was set for 11 a.m., October 11, 1968. There was a short hold in the countdown, and the Saturn 1B belched fire and moved upward at 11:02:45 a.m.

"It's a little bumpy," Schirra reported in the first minute, but then the booster settled down for the fast climb to orbit and Schirra changed his evaluation. "She's riding like a dream."

For the next eleven days, Apollo 7 was both a dream and a nightmare. A few days before launch, Schirra went duck hunting in a Florida swamp with Paul Haney and Jim Rathman, a former Indy race driver and owner of a local Chevy dealership. Both Schirra and Haney came down with colds. Schirra's made him miserable in orbit, and there were days when he was purely grumpy in response to directions from mission control. Haney was *in* mission control eight hours a day, but his interaction was with the news media, and reporters tended to attribute his red eyes to the hours in the press club after he got off work.

The flare-up came twenty-four hours into the mission when Schirra was to unpack a black-and-white TV camera and do live show-and-tell for the waiting world. Even having a TV camera on board had been an internal controversy. Astronauts worried that it was an unnecessary intrusion on their lives. Deke Slayton sent a memo to Chris Kraft way back in 1966 insisting that Apollo planners delete television. Kraft shot a copy of the memo back with a handwritten note:

Deke Slayton—
I can't conceive of this country "sending" 3 men to the moon and not being allowed to see the lunar surface and the sight of a U.S. LEM on the moon. I believe you should reconsider your point of view.

                                                                    Chris Kraft

Slayton was no dummy. He took the proverbial second and a half to reconsider, then called Kraft to say that television would stay on Apollo.

Now the time had come and Wally Schirra was balking. Mission control had already added numerous chores to the crew's work list, they were getting ready to rendezvous with the S-IVB stage that had followed them into orbit, and Schirra wasn't feeling good anyway. "You have added two burns to this flight [of the service module engine], you have added a urine water dump, and we have a new vehicle up here," he radioed querulously, "and I tell you, Flight, TV will be delayed without further discussion until after the rendezvous."

Slayton was in mission control and he flipped a switch on the Capcom console to shift the comm line to his microphone. "All we have agreed to on this particular pass is to flip the switch on," Slayton said. His tone said, *No nonsense, Wally.* "I think we are still obligated to do that."

Schirra was adamant. "We do not have the equipment out," he said. "We have not had an opportunity . . . we have not eaten at this point, and I still have a cold. I refuse to foul up our timelines this way."

Slayton was incensed, but knew that there were two ways to look at it. It was either insubordination or a commander exercising the prerogative to make in-flight decisions. He bit his tongue and let Wally carry the argument.

Mickey Kapp had finally been allowed to provide music tapes to be carried on board, and Schirra listened during quiet times. A few days later, the rendezvous a success and things settling down, he brought out the camera and put on a show. They were 150 miles up when Schirra held up a card provided by Kapp that got a laugh heard around the world: HELLO FROM THE LOVELY APOLLO ROOM HIGH ATOP EVERYTHING.

The camera passed back and forth between Cunningham and Eisele, showing the relatively spacious interior of the command module. Then as they closed the show for the day, Schirra held up another of Kapp's cards: KEEP THOSE CARDS AND LETTERS COMING IN FOLKS.

On another day, Schirra got even with Slayton. The Kapp card he held up read, DEKE SLAYTON, ARE YOU A TURTLE?

The rules of the Intersteller Association of Turtles, Outer Division, require the answer to be "You bet your sweet ass I am." If the respondent is too embarrassed to give the correct reply, he owes everyone within earshot a drink of their choice. Slayton had asked the question during Schirra's Mercury flight; Wally recorded the proper answer on tape. Now Slayton in mission control did the same thing.

"I have recorded my answer," he radioed to Schirra.

When he held up another card for Paul Haney, the public affairs man forgot to record his answer. "Somebody tells me he isn't talking, but just buying," said Capcom Jack Swigert.

"He's buying," Schirra said happily. "Thank you very much."

Haney received mail for years from all over the world, turtle club members reminding him that he owed them a drink. "There must have been 10 million people within earshot that day," he said, "and I think I heard from half of them."

The mission ended October 22. Schirra was feeling better, but now his crewmates had caught his cold, and between them, they'd depleted the onboard supply of antihistamines. It took only a ten-second burn of their big service module engine to kick down out of orbit.

"We're flying a pink cloud," Schirra radioed as they fireballed through the atmosphere. He put the craft down only one-third of a mile from its target point, with the carrier *Essex* standing by to pick them up.

Despite the colds and the grumpy days, it was the best-run American space mission ever—"101 percent of the planned test objectives," George Low said.

When Sam Phillips told reporters that a final decision on the next mission would be made in mid-November, *Time* correspondents in Houston sent this message to the editors in New York:

"Unless something totally unexpected and unbelievably startling is discovered in the mountain of flight data, the decision almost certainly will be to send Apollo 8 and its crew . . . to lunar orbit."

So it was.

13

While America looked toward a Christmas moon, two things briefly took the spotlight away from Apollo 8. Vasily Mishin put Russia back into the manned spaceflight business, and rumors flew that a moon flight was about to follow. Mishin's team sent a new unmanned ship into orbit on October 25, three days after Apollo 7 splashed down, and Radio Moscow called it a Kosmos satellite. Mishin said later that it was an unmanned Soyuz. The next day, Mishin launched cosmonaut Georgy Beregovoy alone in Soyuz 3 in an attempt to rendezvous and dock.

Soyuz was a commodious two-cabin spacecraft. One part was bullet-shaped, a sort of command module with a large cylindrical module behind that carried deployable solar wings, maneuvering rockets, retrorockets, and consumables like fuel, oxygen/nitrogen tanks, and water. Nested to the nose of the control cabin was a huge spherical compartment, so large that cosmonauts could float freely inside. It was a work module, or scientific compartment, voluminous enough to be a full-fledged space laboratory. At its top were an air lock and docking apparatus.

Beregovoy brought the maneuverable Soyuz 3 to within 650 feet of the Soyuz 2 target. That was as good as he could do with the fuel he had. He couldn't lockstep into station-keeping as Wally Schirra had done on Gemini VI and again only thirteen days before on Apollo 7. Docking was out of the question. The cosmonaut stayed aloft four days, then brought his Soyuz down.

But the mission raised new fears that the Russians could be making a move on the moon. They'd missed on several missions, but in the last year had sent two so-called Zond spacecraft out around the moon and brought them home to a safe landing in Russia. Speculation raged that Zond was a cover name for a manned lunar spacecraft and that the Russians would again try to steal America's thunder by sending one man out to the moon and back before Apollo 8, or at least before astronauts could land on the moon in 1969.

Thomas Paine, about to become NASA's third administrator, said that would be a surprise, "a bold step indeed." NASA officials were privy to intelligence reports, but none of them understood that Soyuz was not a lunar ship. As the Russians had been hinting, sometimes forthrightly, Soyuz was a precursor to a space station. Their lunar program was something else and it was nowhere close to beating Americans to the moon.

There was, U.S. intelligence knew, a huge Russian space booster in the offing. It was the long-fabled N-1, with about 10 million pounds of thrust. That made it one-third bigger than the American Saturn 5. Rumors floated from the middle of 1968 on that the N-1 would soon be tested. It didn't happen. But the concern was enough to lead *Time* into an early December cover story titled "Race for the Moon." The cover art showed a space-suited cosmonaut and an astronaut stretching their legs across space in a final spurt to the waiting moon. The artist cleverly chose his perspective so that it was impossible to tell who was ahead.

The Russians had slyly never talked about their lunar spacecraft. So *Time* assumed that Soyuz was it. "Tass, the Russian news agency, has confirmed that both Zonds were preparatory shots for a manned flight," *Time* said in its space race cover story, "and carried living creatures to test radiation effects near the moon. U.S. scientists suspect that Cosmonaut Georgy Beregovoy successfully tested life-support systems for a manned lunar mission during the earth-orbit flight of Soyuz-3. If so, a Soviet lunar spacecraft may finally be man-rated—ready to carry passengers to the moon in December."

It was a good guess, based on the best information that could be gleaned from *Time*'s private sources and from a reading of available Soviet literature. But it wasn't right.

Nobody but the Russians knew exactly what they were up to. And they weren't talking.

It was before dawn on December 24. Christmas Eve. Paul Haney was on the public affairs console in mission control, trying to stay calm as he fed status reports to the newsroom and, he knew, to a worldwide radio and television audience estimated at well over 1 billion people.

Out there behind the moon, if the first-ever rocket maneuver called LOI, *lunar orbit insertion,* had worked properly, three American astronauts were coasting toward history. If LOI hadn't worked, well . . . what was about to happen was anybody's guess. Haney cleared his throat and pressed his mike switch. "This is Apollo control, Houston. Mark. Three minutes from predicted time of acquisition."

That was the moment the Apollo 8 spacecraft would emerge from behind the moon and its radio signals would be picked up on Earth by the expanded worldwide tracking network. "Standing by."

There wasn't much else for Haney to say.

"Apollo control, Houston. Mark. Two minutes . . ."

"Mark. One minute. Jerry Carr [the Capcom sitting one row down from Haney's back-row perch] has placed a call. We are standing by."

Anchormen on television tried to fill the time with something, technobabble, anything. Most reporters in the newsroom were silent, straining to hear what Haney would say next. His voice had a catch in it.

"We've acquired signal but not voice contact yet. We are standing by . . . We are looking at engine data, and it looks good, tank pressure looks good. We have not talked yet with the crew, but we are standing by."

Haney was listening to the flight director's loop in one ear, the Capcom loop in the other. He heard somebody tell Glynn Lunney, tonight's Flight, "We've got it!," and then he heard Jim Lovell's voice. Haney was himself yelling. "We got it! We got it! Apollo 8 now in lunar orbit. There is a cheer in this room!"

He switched to let Lovell's voice go out to the world. ". . . 9 by 60.5," Lovell said. "Good to hear your voice."

Apollo 8 was in orbit around the moon, 169 by 60.5 nautical miles. After two orbits like that, they'd fire their engine again to circularize at 60 nautical miles. This was not apogee and perigee as at Earth. This was apo*lune* and peri*lune.* New ways of thinking were required.

For fifteen frustrating minutes, the world listened to the astronauts and mission control trading engineering information and data on spacecraft performance. Finally they had a few moments to look around. "What does the ole moon look like from sixty miles?" Jerry Carr asked.

Jim Lovell responded. "The moon is essentially gray, no color. Looks like plaster of paris or sort of a grayish deep sand . . . Langrenus is quite a huge crater. It's got a central cone to it. The walls of the crater are terraced, about six or seven different terraces on the way."

After a million years of looking at the moon from 240,000 miles away, man was just 60 nautical miles from the lunar surface. The descriptions went on and Frank Borman, Jim Lovell, and Bill Anders had obviously done their homework. They called off crater names like they were reading street signs back home. ". . . coming up on the craters Columbo and Gutenberg," Anders said with awe. "We can see the long parallel faults of Gaudibert, and they run through the mare material right into the highland material."

They would stay in lunar orbit for twenty hours, ten times around the moon, and have plenty of time for sight-seeing and photography. What mattered was that they were there, that everything had worked, and that this mission all but opened the door for a lunar landing. Nobody had time to think about the Russians and to wonder what leg of the space race they might still be running. On Christmas Eve 1968, nobody cared.

"Earthshine is about as expected, Houston," Bill Anders offered, and it took a moment for the word to sink in. *Earthshine!* It was another new concept to Earth dwellers.

Frank Borman was a religious man and a lay reader in the Episcopal Church. He dedicated a prayer to his church in Seabrook, Texas:

"Give us, O God, the vision which can see thy love in the world in spite of human failure. Give us the faith to trust thy goodness in spite of our ignorance and weakness. Give us the knowledge that we may continue to pray with understanding hearts. And show us what each one of us can do to set forward the coming of the day of universal peace. Amen."

On their third turn around the moon, they were pointing just right to see something that would still be having an impact on mankind long after these adventurers retired to some front porch rocking chair. Bill Anders grabbed a camera and triggered off a series of 70-mm color shots of a blue and green and brown and white ball coming up from the distant moonscape horizon. *Earthrise!*

The dramatic scene left them speechless. The photos became the symbol of their mission, but far more important, *Earthrise* awakened a new awareness of one small planet's fragile place in a dark universe and became the enduring symbol of generations of environmentalists.

Frank Borman curled up on his couch, a mesh sleeping bag keeping him from floating away. He'd been sick in the early part of the mission, vomiting and with loose bowels. Reporters who heard this and who knew of the infamous Gemini V simulation when the script had the air in the cabin turning into a brown haze wondered about conditions aboard Apollo 8. But the waste management system had contained things, and lithium hydroxide canisters in the environmental system eventually filtered out some unpleasant odors. Flying in space was not all glamour. Borman and Lovell knew the downside after their fourteen days together aboard Gemini VII. Bill Anders learned a few home truths on the way to the moon.

So Borman slept while Lovell and Anders, too stoked on adrenaline to even catnap, tracked landmarks, took pictures, and checked out Apollo's navigation system. Their experience would help future crews. Finally Borman awoke and he ordered—he had to order—the others to get some sleep. They did, but only a few hours. Then they were up and at the windows again.

It was dark now in Houston, Christmas Eve when families gathered around tables and decorated trees, children couldn't imagine sleeping because Santa Claus was coming, and the grown-ups drank a toddy and put holiday music on the record player. But not *this* Christmas Eve. At thousands of homes around Clear Lake and the Manned Spacecraft Center, Dad or Mom or both were at work. Christmas Eve baby-sitters, a mythical breed in normal times, tuned television sets to any channel at all, and the kids at home put Santa aside for an hour to watch something they didn't quite understand, but were told was important.

At the space center newsroom, reporters who were far from home for Christmas waited for the story that would be the most important they'd ever covered. For a bunch of allegedly hard-bitten types, there were a lot of lumps beginning to grow in a lot of throats.

The anticipation was not confined to Clear Lake. Around that multicolored Earth, a television audience estimated at close to 1 billion people waited too. Christmas Eve in 1968 was anything but traditional. Just after 8 p.m. Houston time, technicians in mission control set their switches, and a live television feed went out to the world. The first images were of a stark black-and-white cratered moonscape. The voice was Frank Borman's.

"This is Apollo 8, coming to you live from the moon."

A quarter million miles away, a world stopped to watch and listen. Each of them described what he was seeing, what he felt. To Earthlings, the words were almost unheard; it was the moonscape that fascinated. It went on for nearly twenty minutes before Bill Anders began the Christmas message that Frank Borman had brought along. It wasn't just Borman's message; it came from antiquity and from three modern astronauts orbiting Earth's moon.

"In the beginning," Anders read, "God created the heaven and the earth . . ." Lovell read next, and finally Frank Borman. Apollo 8 was approaching the horizon and would soon pass behind the moon again.

". . . and the gathering together of the waters called He seas: and God saw that it was good." The other side of the moon was coming fast. Frank Borman closed out a Christmas Eve that many on Earth would never even think about forgetting. ". . . from the crew of Apollo 8, we close with: Good night. Good luck. A Merry Christmas. And God bless all of you," he said, "all of you on the good Earth." If there were dry eyes, they were not in the news center, where reporters had forgotten to take notes.

They had Christmas dinner the next day, on the way home. The TEI burn on the back side of the moon—trans-Earth injection—was perfect, and when they came around the moon that time, they were no longer in lunar orbit. They were heading back to Earth. As they emerged and mission control sighed in relief, Lovell radioed to Earth: "Please be informed there is a Santa Claus."

Dinner was a surprise. They'd been expecting freeze-dried again. They got real turkey with dressing and cranberry sauce in the same kind of pack that troops in Vietnam were eating that day in the killing fields. That part didn't occur to them; what they liked was the taste of turkey after all that freeze-dried.

Tucked in with dinner was a surprise from Deke Slayton: three minis of brandy for a trans-Earth toast. Frank Borman took a stern look and said three words. "Put them back." Years later the minis sat in proud display, still unopened, on Bob Gilruth's fireplace mantel in Dickinson, Texas.

Apollo 8's parachutes opened over the top of the *Yorktown* waiting for them in the Pacific and touched water less than three miles away. The world watched again on live television. It wasn't a great year in American history. The Tet Offensive was a disaster for the Vietcong, but somehow they made it look like a victory. Lyndon Johnson threw in the towel over Vietnam, student riots, and national malaise. Martin Luther King and Bobby Kennedy were murdered. More riots marred the Democratic National Convention in Chicago. Richard Nixon won the White House.

It wasn't a great year in American history until Christmas Eve. Then NASA and Apollo saved 1968. *Time* put Borman, Anders, and Lovell on its cover as Men of the Year.

The last lap of the space race was about to begin.

Vasily Mishin put Soyuz 4 into orbit January 14, 1969, with Vladimir Shatalov aboard. The next day he launched Yevgeny Khrunov, Aleksei Yeliseyev, and Boris Volynov in Soyuz 5.

The three-man ship was in a slightly higher orbit, so Shatalov caught up with them. His rendezvous maneuvers were just right for station-keeping. Then while Soyuz 5 stayed passive, Shatalov plugged 4's docking mechanism into it and the two ships were joined.

With the laboratory modules locked together, it was another first for the Russians, two spacecraft coming together to create a legitimate, though relatively small, space station. They played it for all the propaganda value possible. But they weren't done. After Shatalov backed away, Khrunov and Yeliseyev did a two-man space walk and moved permanently into Soyuz 4. It was an impressive demonstration of how an orbital emergency might be handled. So now they had a crew transfer in space to add to their list. Shatalov and his two new crewmates landed January 17, and Boris Volynov came home alone a day later.

There was a brief flurry of news speculation about the Soyuz flights foreshadowing some kind of Russian moon mission. Then the growing expertise of the space press corps prevailed. Soyuz was almost certainly aiming for something other than the moon. If they wanted to put up a space station, so be it.

Still, it wouldn't hurt to keep one eye on the Russians while the Americans got ready for the final rush to the lunar surface.

Deke Slayton called Neil Armstrong into his office in January and gave him the Apollo 11 mission, the one likely to be first on the moon. His crew was Buzz Aldrin and Mike Collins. Armstrong and Aldrin had been backups on Apollo 8. In Deke's way of doing things, backup crews rotated ahead three missions to become prime crews. Collins was fresh on duty after recovering from his back surgery. He'd missed the chance to go to the moon on Apollo 8. Now he regained his position in the rotation and was the command module pilot on a mission that might be even more important.

Buzz Aldrin was delighted with his assignment as lunar module pilot (LMP). Everyone knew that the lunar landing flight plan called for the LMP to be the first man to set foot on the moon. If Apollos 9 and 10 worked right, Edwin E. "Buzz" Aldrin's name would forever be the answer to a question on school kids' history tests.

But 9 and 10 had to be right. Jim McDivitt, Dave Scott, and Rusty Schweickart were in low Earth orbit—the D mission—on March 3, 1969. McDivitt was the crew commander and Scott was the command module pilot. Schweickart was lunar module pilot. McDivitt's and Scott's job titles reflected their real assignment. McDivitt was the overall boss and Scott controlled the command module in its maneuvers. Schweickart was in a different category. The LMP was really a copilot; when the time came, flying the lunar module was the commander's job. The LMP was his assistant and wouldn't get to touch the controls except in an emergency that disabled the commander.

The first chore in orbit was Scott's, and it had never been done before. He pulled away from the S-IVB rocket stage and turned the command and service modules around. The sight in front of them was awesome. The four adapter panels of the "garage" had blown away, and on top of the big rocket sat their lunar module. Gold foil to reflect solar heat and maintain temperature equilibriums glittered against the LM's black, angular structures.

Quickly there would be two manned spacecraft in orbit and the radio call sign "Apollo 9" wouldn't work. From now on, astronauts again were allowed to name their ships, and if a few joyless souls at NASA headquarters winced, they didn't send memos about frivolous names. The Apollo 9 ships were named for their shapes. The command module was *Gumdrop.* The weird-looking lunar module was *Spider.*

With *Gumdrop*'s pointy-nosed docking probe exposed, Scott blipped his thrusters and glided toward the funnel-shaped drogue in *Spider*'s roof. It took two tries to get it right, but finally they heard a satisfying clunk and clicks as the probe locked into the drogue and the mechanism pulled the two ships tightly together. Lights and instruments on *Gumdrop*'s panel verified that they were captured and locked. Eventually they plugged together cables to provide power to *Spider,* finished a checkout, and then Scott reversed thrust and pulled *Spider* from its garage and extended its folded legs. He'd just duplicated the Russian feat with Soyuz six weeks earlier. Now the Apollo 9 crew planned to do things that the cosmonauts could only imagine.

By the end of the day, Scott had used the service module's big engine several times to move the combined ships up to a 222-by-316-mile orbit. All three had experienced some dizziness as they adjusted to zero gravity. Rusty Schweickart didn't adjust; he got worse, developing a severe case of vomiting space sickness. It slowed him down, but the work went on. The vomiting would come on suddenly, then he'd feel better.

Over the next several days, they opened the hatches in *Gumdrop*'s nose and *Spider*'s roof and checked out *Spider*'s interior. McDivitt fired up *Spider*'s big descent engine, and they put another check mark on the long list that would lead to another crew landing on the moon. But Schweickart's sickness put a crimp in the next exercise. He was supposed to don a backpack similar to the one astronauts would wear on the moon. Then with both *Gumdrop* and *Spider* depressurized, he'd go out *Spider*'s forward hatch and move across to *Gumdrop*.

It was just what the Russians had done on the January Soyuz flight. McDivitt said no. When Schweickart seemed to be feeling better, they found another way to do 90 percent of the exercise. With cabins reduced to a vacuum and their space suits pressurized, Schweickart opened *Spider*'s hatch and went out onto the porch. Scott opened *Gumdrop*'s hatch and stood up. Both were secured by tethers; they didn't want to accidentally float away. That was enough. Nobody doubted that Schweickart could have crossed to *Gumdrop* and gotten inside. The points of the exercise—depressurize both craft and do a rescue—were satisfied.

All that was left was the big one. *Spider* had to go off on its own with McDivitt and Schweickart, leaving Scott alone in *Gumdrop*. They did it March 7, undocking and getting as far as 115 miles apart. This was the risky time. If *Spider*'s equipment failed, it couldn't reenter the atmosphere without burning up. It had no heat shield and its shape was all wrong. It was perfect for landing on the moon, terrible for landing on Earth. The backup plan was for Scott to go get them with *Gumdrop*. The same kind of plan would apply in lunar orbit. On Apollo 9, and ever after, the plan was never needed. *Spider* was solo for six hours and did everything McDivitt asked of it. When *Spider* and *Gumdrop* docked again—"We have capture," McDivitt told mission control—nine astronauts were proud, pleased, and happy. The other six were in Houston, where the rendezvous and second docking had just locked in their plans for the summer of '69.

Bob Gilruth confirmed their surmise with only the slightest born-of-experience hedge in one of his increasingly rare out-front appearances. "When all the data are in," he told reporters, "I believe we'll be set for the final flights to a lunar landing."

The static was getting bad and Apollo 10 commander Tom Stafford was getting frustrated. "Aw, this fucking comm," he groaned, just in time for the static to clear and his words to be heard a quarter million miles away.

"Houston, this is *Snoopy*." The voice now was that of Gene Cernan, the LMP, and he was elated. "We is *go* and we is down among 'em, Charlie."

Charlie Brown and *Snoopy* had gone to the moon. Now *Snoopy*, the lunar module, was dropping down to just 47,000 feet above the surface. Tom Stafford and Gene Cernan couldn't begin to contain their excitement. High above in a sixty-nine-mile circular orbit, the command module, *Charlie Brown*, was passing and John Young was wishing that he was down there among 'em too.

The Apollo E mission fell off the books after Christmas. This was F—take the ships to the moon and do everything with them except land. Tom Stafford was tempted to disobey and put *Snoopy* down anyway. His discipline as an Air Force officer and the fact that *Snoopy* was over the allowable weight limits held off his urge. But the next time somebody got this close to the moon, his LM would be lighter and he'd be cleared to land.

They were coming up on one of the possible landing sites when Stafford's camera jammed. "This goddamn filter has failed me," he muttered loud enough to be heard in Houston. "My Hasselblad just failed."

In the mission control view room, one old-time NASA official turned to another and grinned. "God," he said, "isn't it good to hear men talking up there instead of boys?"

It quickly got better. One of the obvious landmarks as they swooped down to only nine miles above the moon was a big crater. Stafford spotted it first: "I've got Censorinus right here bigger than shit!"

"Son of a bitch!" Cernan yelled more than once. At 47,000 feet, not much higher than most airliners fly, the craters and mountains of the moon were all around them. "Goddamn! That one looked like it was coming inside."

As they skimmed the moon, Cernan was elated. "It looks like we're getting so close all you have to do is put your tail wheel down and you're there."

Then it was time to go back up to *Charlie Brown*. Stafford aligned *Snoopy* and got ready to drop off the descent stage. Then he'd fire the ascent engine that would kick them up toward the first rendezvous in lunar orbit.

"You ready . . . son of a bitch!" Cernan hollered. *Snoopy* was suddenly flinging itself in every direction at once. Their stable, flyable lunar module had gone crazy. "Let's make this burn on the AGS, babe. Make this burn on the AGS."

AGS was the abort guidance system. It was for emergencies only. A switch setting had been left off their checklist. The firing descent engine had snapped to the left and almost locked. Stafford grabbed the thrusters, held *Snoopy* steady, and they staged away leaving the bottom half of *Snoopy* behind. Now under power of the ascent engine, Stafford guided them smoothly toward rendezvous.

"Looks like we got a good staging out of it," Stafford said, a measure of calm in his voice. "We're all set for insertion here."

"I'll tell you, that was wild, babe," Cernan gritted. It had been a near thing, saved by their quick reflexes and training.

". . . but not dangerous," flight director Glynn Lunney said later. "It simply rotated the vehicle . . . the machine did something unexpected." Nor would it change the plans to let Apollo 11 try for a landing.

The rest was as routine as something being done for the first time can be. "*Snoopy* and *Charlie Brown* are hugging each other," Stafford radioed after John Young flew the command module the last few feet to a docking. They crawled back to join Young, dumped *Snoopy,* and got ready to go back to Earth.

Then just as it was getting to be too routine, Gene Cernan's report woke them up in Houston. "The crew status is tired and happy and hungry and thirsty and horny, and all those other things," Cernan chuckled.

Horny?

With Deke Slayton's permission, the flight plan carried in *Charlie Brown* had been spiced up a bit. There was a centerfold from the May 1969 *Playboy* stapled into the book just before the pages describing lunar orbit insertion. Every couple of pages, there was more. When the flight got to Apollo 10 going around the back of the moon, there were naked female breasts framed in "left turn" and "right turn" road signs.

At rendezvous, John Young had flipped a page and found a road sign depicting spread female legs. "Divided Pavement Ends" the legend read. Within the next few minutes, his job was to insert *Charlie Brown*'s probe into *Snoopy*'s drogue. He managed it without laughing.

"We just didn't want you to forget what the good things are like back on Earth, Gene," radioed Capcom Joe Engle.

"How can I? I keep looking at this flight plan."

In Houston, Buzz Aldrin was looking at a flight plan too, and not liking what he saw.

The rumor, widespread enough so that it appeared in newspapers, was that Neil Armstrong exercised his commander's right to decide who did what. His purported decision was that Buzz Aldrin would be the second man on the moon. The rumor was wrong. A few weeks after they began training, flight planners made a change and it was approved all the way up to Bob Gilruth. But the rumor persisted and even the crew didn't know quite what had happened. In his book *Carrying the Fire,* Collins still said that Armstrong was responsible. He wasn't. It was Deke Slayton.

"For the two-man EVA," the first paragraph of the summary plan now read, "the CDR will always egress first and ingress last to maintain proper LM crew station configuration."

Translation: The lunar module forward hatch was hinged on the right. When it opened, it blocked the LMP (Aldrin). For him to get out first, he'd either have to crawl around the hatch in a pressurized space suit and wearing a big life-support backpack, or he'd have to crowd in on Neil Armstrong before they opened the hatch.

Slayton had analyzed the situation through his engineer's eyes and realized that it was virtually impossible to do either. There was only one way out. The commander would crawl out first. Later, the LMP would push the hatch mostly closed, move to the other side of the cabin, open the hatch, and go out himself. Getting back in required the same steps in reverse.

Buzz Aldrin didn't like it. In Gemini, it was the right-seater who got out while the commander stayed put. He'd gone out himself on Gemini XII while Jim Lovell stayed in his commander's left seat. He argued the point, convinced that Armstrong had personally done him a bad deed by usurping his chance to make ultimate history. "No, it wasn't that way," Armstrong said whenever he was asked, and the question would come up again and again forever. "It wasn't my decision. It was a procedures decision."

But it left a bitter feeling in Aldrin's heart; he and Neil Armstrong would never be more than passing acquaintances after that.

Apollo was rushing toward the Kennedy deadline, flying missions almost as fast as Gemini had—September, December, March, May, and now the ultimate mission in July 1969. At odd moments, somebody would look around and ask, "Where are the Russians?" They'd been there from the beginning, obvious and gloating for years, more recently fading into a shadowy presence that left the hairs on the backs of NASA necks sometimes rising for fear that somehow the Russians would do it again.

They tried.

In early July, the monstrous N-1 rocket with 10 million pounds of thrust was in final checkout for launch. It was a moon rocket with no other purpose, but there were no men in the prototype lunar craft at its tip. It would be twenty years before the Soviet government even admitted that the rocket existed.

Whatever went wrong was catastrophic. The N-1 exploded with a force that destroyed its Baikonur launch facility—pads, supporting structures, everything. More than one hundred people died, and the damage was so great that it would take two years to rebuild. In one horrible accident, the Russian manned lunar landing program went into a downward spiral from which it never recovered. American intelligence had been watching the N-1 preparations with unease, diverting ever-better spy satellites to the region and analyzing high-quality photos on a daily basis. One set of photos showed the rocket there. The next set showed blackened devastation. NASA officials were quietly informed. The Russian threat, they assumed, was gone.

Not quite. Vasily Mishin had one last mission. He'd long known that no cosmonaut could land on the moon ahead of the Americans. But if the first lunar sample came home to Mother Russia, that would be a major victory. He launched Luna 15 on July 13, 1969. Three days later, the same day Apollo 11 was launching from Kennedy Space Center, it braked into lunar orbit. The Russian space team was taking its last, best shot. Luna 15's mission was to land on the moon, scoop up a sample of moon dirt, then blast off and return to Earth.

Moon samples would be in Russian science laboratories long before anything came home from Apollo 11. Twice Mishin's control team sent commands to lower Luna 15's orbit and it responded. Quiet back-channel communications between the American State Department and its Soviet counterpart brought assurances: Luna 15 in no way intended to interfere with Apollo 11.

Its assignment at the moon, though, was obvious. American officials quietly worried that it would land, scoop up some dirt, and bring it back to Earth before Apollo 11 could complete its mission. Moon rocks in Russia next week? The thought was chilling.

At midafternoon on July 20, 1969, the command module *Columbia* and the lunar module *Eagle* were separately on the back side of the moon. *Columbia*, with Mike Collins watching his crewmates fall away, was in a stable circular orbit. *Eagle* was descending.

The long race was on its last half-lap.

On a lawn outside the Manned Spacecraft Center, a loud group of civil rights protesters had gathered. The money being wasted on this lunar landing should be spent helping the poor and downtrodden in America. When Mickey Kapp heard that they were singing the songs of protest, he grabbed a pro-model tape recorder and went out to catch the action for a record album set he was producing for Time-Life Records.

Just before he got there, the small crowd was singing "He's got the whole world in his hands . . ." Now that *Eagle* was on the way down, the mood had changed. They were singing, but there was a different passion in the lyrics.

"He's got the astronauts in his hands,
"He's got the astronauts in his hands . . ."

Behind the moon, *Eagle*'s legs were extended and pointed slightly forward. Neil Armstrong and Buzz Aldrin could see the moon below, but they were facedown going backward. What they saw was what they were just passing.

Around to the front of the moon, now with Houston watching the stream of data and in voice communication, Armstrong and Aldrin didn't quite hear the words from the Capcom, Charlie Duke.

"Eagle, Houston. If you read, you're a go for powered descent. Over."

After a silence, Mike Collins relayed the message. *"Eagle,* this is *Columbia.* They just gave you a 'go' for powered descent."

Eagle's antenna was pointing just a little bit off. They couldn't hear Duke, and Houston couldn't hear them. Duke asked Collins to tell Armstrong to yaw sideways so the antenna would lock on. The signals strengthened.

"*Eagle*, Houston. We read you now. You're go for PDI" (powered descent initiation).

Armstrong and Aldrin watched their instruments, particularly the computer readouts. *Eagle*'s radar, combined with its navigation software, would tell them where they were, where they were going, and how fast. They were go to fire their braking rocket, and at just the right second, Aldrin hit the button.

"Ignition." *Eagle*'s descent engine fired up and Armstrong had control. They were ten minutes from landing. Aldrin timed the passing checkpoints, letting Armstrong focus on the automated systems holding *Eagle* at the right attitude and running the engine. The landmarks—craters, craggy mounds, tracks in the lunar dirt made by rolling chunks of ejecta from the impacts that made the craters—zipped past just a bit early. If that held, they'd land long.

Aldrin ran the computer. When he set it to check altitude data, a buzzing alarm went off.

"Program alarm," Armstrong radioed. He was quick; reporting alarms was a reflex he'd acquired in months of training in the Apollo simulators. "1202. 1202."

That was the code that flashed on the computer's readout. Neither of them knew what it meant. A 1202 had never showed up in training exercises.

In mission control, the guidance office—GUIDO on the voice loop—was a twenty-seven-year-old named Steve Bales. He knew what it meant. The computer was overloaded and was simply recycling to digest the incoming data. Gene Kranz was flight director. He was looking up, but he didn't have to ask. "We're go on that, Flight," Bales said.

Charlie Duke, like all of them, monitored Flight's voice loop. He didn't wait either. "*Eagle*, Houston. We're go on that alarm." Aldrin punched the right computer buttons and *Eagle* had data.

It happened again. "1202. 1202." The buzzing master alarm was irritating and what if it meant something bad?

"We're go, *Eagle*," Duke called. The irritating alarm sounded three more times while they braked rapidly toward the moon.

Bales was on it instantly. He was fifteen years old when the Chief Designer sent Sputnik 1 into orbit. Now he was the voice of calm and reason in mission control. "We're go, Flight," and Charlie Duke passed the word to Armstrong and Aldrin.

Suddenly they weren't that high anymore. At 7,500 feet, *"high gate,"* the lunar module automatically pitched up and over. They could see where they were going. The descent engine had throttled up, and *Eagle's* slowdown was obvious. In Houston, they were watching the same data that Aldrin had, only more of it and, because of the distances involved, one and a half seconds later. It looked good.

"You're go for landing," Charlie Duke called. The men aboard *Eagle* heard, but didn't acknowledge. They were busy.

They were below 3,000 feet, approaching *"low gate,"* when the alarm buzzed. It was different. "Program alarm," Aldrin reported. "1201."

Same thing, another kind of overload, Steve Bales said. "We're go," Charlie Duke radioed.

They were holding their breath in the Armstrong house in El Lago. Jan Armstrong watched on television—there was no live TV of the landing, but the networks did it nicely in animation—and listened to her squawk box. This time she knew that no matter what happened, it would not be turned off. The Gemini VIII experience would not be repeated. A few weeks earlier she confronted Deke Slayton at the last social gathering before the guys went to the Cape for the final preparations. Deke was not a man who took kindly to being *told* anything. There were a lot of days when he didn't even like to be *asked.*

Jan Armstrong told him.

"This time," she said, "don't you even think about cutting me off. If something happens and you cut me off, the whole world will know about it." There'd be a press conference on the front step, and the angry lady doing the talking would be Jan Armstrong.

Eagle was gliding across the moon, barely above it, and Armstrong could see that they were heading for a boulder field. It was not a good place to land.

Three hundred feet. Armstrong took partial control, tilting left, adjusting the angles so they passed the boulders, came up on a big crater. That was no good either.

Buzz Aldrin was calling out numbers, how high they were, how fast they were moving in feet per second. "Sixty feet. Down two and a half. Two forward . . . two forward."

They were crawling now, just barely above the surface. Houston was watching their fuel consumption. "Sixty seconds," Charlie Duke radioed. Just a minute of fuel left. Neil Armstrong picked his spot. He could get there and down, no boulders, no steep crater walls.

"Getting some dust here." It was blowing up from the rocket thrust. They had to be close. There was a metal probe sticking down from a landing pad. When it touched the moon, a signal light in *Eagle* would come on.

"Contact light," Aldrin said.

"Engine stop."

"Engine arm off."

Aldrin's staccato reports flashed back to Houston. Then there was a long moment of silence. The next voice from *Eagle* was Armstrong's, and the next words were spoken from the surface of the moon. It was 3:17 p.m. Houston time, July 20, 1969.

"Houston, Tranquillity Base here. The *Eagle* has landed."

"Roger, Tranquillity," Charlie Duke shot back. "We copy you on the ground. You got a bunch of guys about to turn blue. We're breathing again. Thanks a lot."

The Finish Line

It was midevening the same day, almost 9:30 p.m., in Houston. The next day Luna 15 would crash into the moon and be destroyed. At this moment, Neil Armstrong and Buzz Aldrin stood awkwardly in their pressurized space suits, looking out *Eagle*'s triangular windows at a gray and dusty landscape.

Four days ago they were looking at blue sky and green palmetto trees and white foam on the beach surf. Fourteen years ago, a man in Russia that none of them knew started them on this journey to Tranquillity Base. Sergei Korolev, the Chief Designer, challenged the United States to a race. It took a while to start running, but it was impossible for America to say no. Now it was time to finish it.

Oxygen whispered out of *Eagle,* venting to the moon and creating an indoor vacuum. In the next five minutes, Neil Armstrong had to go out there. Then he had to make a comment for the record.

I have no idea what to say.

He raised his eyes to the horizon only six miles away. Between here and there, nothing changed much. *But between* there *and* here, *everything has changed. How did we get from* there *to* here?

He looked down to the ground and it didn't seem so far away. *It's just a small . . .* His mind took the thought and raced to the logical conclusion. *Of course! It's both. It's always been both. That's what I should say.*

Buzz Aldrin tugged on the hatch and it wouldn't open. *Still not a vacuum in here.* He waited a few seconds, tried again, and it slowly gave way.

"Hatch coming open," Neil Armstrong said. On Earth, a world of people waited.

With a nod to Aldrin and an unseen smile, he turned around, crouched, and began to back out. The porch sloped down just a bit. He reached over to a D-ring, pulled it gently—in the one-sixth gravity of the moon, he had the strength of Samson—and a desklike panel in *Eagle*'s side came open. There pointing at Neil Armstrong was a black-and-white TV camera. A fuzzy but awesome image appeared on television sets in every part of planet Earth.

There was this bulky human form in a ballooned white suit, backing down a ladder so very far away. The steps were easy, but the last one was higher than Neil Armstrong expected. He jumped into the big circular dish at the bottom and still hadn't touched ground.

An entire world watched. They listened to him briefly describe what he saw out there on another world. They wished he'd get on with it.

"Okay." His voice was scratchy with distance and static. "I'm going to step off the LM now."

Neil Armstrong put his left foot down into the dust. He felt that the ground was solid underneath. The other foot came down. He took a deep breath and the words he'd written in his mind only minutes before flowed slowly from his mouth and then on to immortal history.

That's one small step for man . . . One giant leap for mankind.

The race was over.

America won.

Epilogue

Bob Gilruth built the manned space-flight team, found the best people he could for key positions, then trusted them to do the job. He made many of the big decisions in the race to the moon, or championed positions until some bureaucrat in Washington made the decision he wanted. He was the boss in Houston, but he let his people meet the press, take the bows, and become heroes.

Jean Gilruth died of a heart attack in early 1973. Bob Gilruth remarried soon after. He and his new wife, Jo, finished work on the boat, christened it *The Outrigger,* and in July, 1973, zigzagged it through the trees in the backyard and launched it into Dickinson Bayou.

By now, his life's work long done in leading Americans to the moon, Gilruth had ceded directorship of the Houston space center to Chris Kraft and taken a job at NASA headquarters in Washington. He and Jo sailed *The Outrigger* through the Gulf of Mexico, up the East Coast, and into Bob's familiar Chesapeake Bay. From there, and for years after he retired from NASA, they sailed the near oceans and looked on clear nights at the stars and the conquered moon. He did not sail around the world, perhaps his only goal unreached.

The other major figure in the American moon program, Wernher von Braun, died in 1977. Of the three giants who ran the race—Korolev, von Braun, and Gilruth—only Bob Gilruth still lives on January 1, 1999. What thoughts flash through his brilliant mind are locked behind the finality of Alzheimer's.

Godspeed, Bob Gilruth.

Index